David M. Hill
Latex Dipping

Also of interest

Shape Memory Polymers
Theory and Application
Kalita, 2018
ISBN 978-3-11-056932-2, e-ISBN 978-3-11-057017-5

Polymer Engineering
Tylkowski, Wieszczycka, Jastrzab (Eds.), 2017
ISBN 978-3-11-046828-1, e-ISBN 978-3-11-046974-5

Polymeric Surfactants
Dispersion Stability and Industrial Applications
Tadros, 2017
ISBN 978-3-11-048722-0, e-ISBN 978-3-11-048728-2

Electrospinning
A Practical Guide to Nanofibers
Agarwal, Burgard, Greiner, Wendorff, 2016
ISBN 978-3-11-033180-6, e-ISBN 978-3-11-033351-0

e-Polymers.
Editor-in-Chief: Seema Agarwal
ISSN 2197-4586
e-ISSN 1618-7229

David M. Hill

Latex Dipping

Science and Technology

2nd Edition

DE GRUYTER

Author
David M. Hill
David Hill & Associates
Elsenhem
Bishop's Stortford
CM22 6LT Hertfordshire

ISBN 978-3-11-063782-3
e-ISBN (PDF) 978-3-11-063809-7
e-ISBN (EPUB) 978-3-11-063823-3

Library of Congress Control Number: 2018966604

Bibliographic information published by the Deutsche Nationalbibliothek
The Deutsche Nationalbibliothek lists this publication in the Deutsche Nationalbibliografie; detailed bibliographic data are available on the Internet at http://dnb.dnb.de.

Typesetting: Integra Software Services Pvt. Ltd.
Printing and binding: CPI books GmbH, Leck
Cover image: lnzyx / iStock / Getty Images Plus / Getty Images

www.degruyter.com

Preface

This book is the product of many years of experience in the production of condoms and gloves. During that time it has become apparent to me that, whilst there are many excellent books on such topics as latices, colloid science, polymer science, vulcanisation of elastomers and so on, there is little coverage of the type of situations frequently encountered on the typical dipping line. This book tries to redress this. It is aimed at supervisors, laboratory staff and technical managers in manufacturing companies producing dipped latex articles, although it is hoped that others may also find it both useful and interesting.

The book covers the principal raw materials that will be encountered in latex dipping and how these are used in compounding the latices. It gives a top-line description of wetting, latex stability and film-forming, and a similar simplified picture of vulcanisation and degradation (primarily for natural rubber). The operations encountered in the different types of latex dipping are described, together with other areas that will be encountered, such as analysis, troubleshooting, process validation and quality control. Finally, an estimate is proposed of the areas where latex dipping may progress in the future.

The book is drawn from my experience in research and development, process development and technical support during 36 years working for a major player in the dipping industry and as a technical consultant. It includes much derived from the many training courses I have run over these years. However, again reflecting my experience, a major focus of the book is on condoms and gloves (surgical, industrial and houseware), with an emphasis on natural rubber latex. As such the reader will find that the level of detail will vary depending on the subject area covered. However, it is hoped that those working with other end products, such as balloons, baby bottle teats and catheters will find that much of the subject matter discussed can be applied to their particular industry as well.

References are given wherever possible, to allow the interested reader to delve more deeply into specific subject areas. It is noted that many of these references refer to work carried out some considerable time ago: it seems that modern research publications, in the main, prefer to look deeper into the science. Interesting and valuable, but maybe not primarily to the target readership of this book.

It has become clear during my experience that the latex industry is a curious mixture of homogeneity and diversity. Many latex formulations, from many companies, turn out to be quite similar. This should not be surprising: companies are making similar products, selling into the same markets, subject to the same regulatory constraints and performance standards, and having to obey the same laws of chemistry and physics. They are using similar raw materials from similar raw material suppliers and have access to the same technical information. However, some of the ways in which production challenges are met are intriguingly different. For this reason, many of the recommendations proposed in this book, and based on my knowledge and

https://doi.org/10.1515/9783110638097-201

experience, may differ from practices being used by the reader. Two pieces of advice here – if your method works well, don't change it; but read, consider and remember the advice. It may serve you well in the future.

It will no doubt be noted by many readers that there is a lack of absolute scientific rigour in some of the subjects tackled within the book. This is an attempt to present the topic in a way that will be easily understood by the target readership. It is hoped that those who know more will forgive any oversimplification. Wherever possible references to more indepth treatments are supplied for those who wish to explore the subject further.

There are occasions within the book where the same subject will be found to be repeated to some extent. It is recognised that there will be many times when the reader may want to refer to a particular subject, and an attempt has been made to contain all relevant information within a section, even if similar information can be found elsewhere in the book.

I am very aware that this is a single book, written by a single author. It is one view of a subject that, like so much in life, reveals more and more complexity the further that one explores it. There will be some who disagree with some of the conclusions, but, as stated, latex compounding and dipping is a complex and multifaceted subject. It is also a fascinating area, which has intrigued, frustrated, but ultimately rewarded me over many years. I hope that the reader experiences the same.

This book is very much a personal view, but I would like to acknowledge the contributions of colleagues over the many years of working in this industry. Space does not permit me to name all, but I would like particularly to acknowledge the many interesting and informative discussions with the late Professor Maurice Porter and Dr Bill Potter.

Finally, two others who played an important part in my early years in the industry, and who gave me much useful advice, guidance and friendship – the late Dr David Blackley and the late Dave Bratby. To them this book is dedicated. I hope that they would have approved.

David Hill
July 2017

Contents

1 Raw materials

1.1 Polymer latex or polymer solution?

In general, polymer latices are the material of choice for dipping, rather than solutions, for the following reasons. Unless the polymer is of a low molecular weight (MW) or at a low concentration, polymer solutions are viscous, which will result in problems with flow and controlling pick-up, and are prone to air being entrained giving rise to bubbles and possibly holes in the finished product. If the solution is diluted to a lower viscosity, the finished rubber film will be thin and several dips will be required to give a useable product. Latices, on the other hand, can be of a high solids content – typically 45 to 60% – coupled with a much lower viscosity. Polymer solutions are usually based on an organic solvent, which can be flammable, toxic or both, whereas latices are water based. It is possible to obtain aqueous solutions of some polymers, but the constraints of solids content and viscosity still apply. Some products are dipped from polymer solutions – for example, polyurethane (PU) condoms and some electricians' gloves – but for the reasons outlined above, the overwhelming majority of dipped elastomeric products are formed by dipping latex. The nomenclature used here follows that described by Blackley [1]. Natural latices are produced *via* metabolic processes in trees and other plants, synthetic latices are produced by emulsion polymerisation of the relevant monomers and artificial latices are produced by dispersing the polymer (however, it may have been produced) in water.

Not all latices, however, are suitable for dipping. The polymer has to be film-forming at the temperatures used in the process – that is, the individual polymer particles which comprise the latex must be able to lose their individual identity and fuse together into a coherent film. The subject of film-forming will be dealt with in more detail in a later chapter, but these requirements mean that there are relatively few latices that are used commercially in dipping. Those in significant use are:

- Natural rubber (NR).
- Acrylonitrile-butadiene rubber ['nitrile' rubber (NBR)] and carboxylated nitrile rubber (XNBR).
- Polychloroprene (CR) or chloroprene rubber. 'Neoprene', the trade name of DuPont's CR, is in general use for this material.
- Styrene-butadiene rubber (SBR) latex, although widely used in applications such as carpet backing and foam rubber, it is rarely used in dipping applications, although it can be incorporated into other latex formulations as a reinforcing filler.

There are also some artificial latices which are suitable for dipping and include:

- Synthetic *cis*-1,4-polyisoprene rubber (IR) – the synthetic version of NR latex.
- PU rubber.

https://doi.org/10.1515/9783110638097-001

- Isoprene-isobutylene rubber (IIR) ('butyl'rubber).
- Ethylene-propylene diene monomer rubber (EPDM).
- Chlorosulfonated polyethylene rubber (CSR).

Of these artificial latices, only the IR latex is currently being used to a significant extent in the dipping industry.

There are many other types of latices but none is used commercially to produce dipped rubber articles, although as mentioned above, SBR latex can be used as a reinforcing filler, as can polymethyl methacrylate (PMMA) latex.

1.1.1 Natural rubber latex

NR latex is a high molecular weight (up to 10^6 daltons) polymer of *cis*-1,4-polyisoprene, produced by the rubber tree *Hevea Brasiliensis*, which as the name implies, is a native of South America. to Kew Gardens in London towards the end of the 19th century. Seeds of the rubber tree were taken by the British explorer Sir Henry Wickham from Brazil Some of the seeds germinated, and seedlings were then transferred to Sri Lanka and Singapore, starting the cultivation of rubber in South- East Asia. Successful cultivation of the tree requires a tropical environment, with high temperatures and high annual rainfall. It is usually grown at lower levels and is rarely seen above 400 m. Intensive study over the years has resulted in the development of many different clones of the original strain, selected for high yield, rapid growth and resistance to disease. The structure of the NR molecule is shown in Figure 1.1.

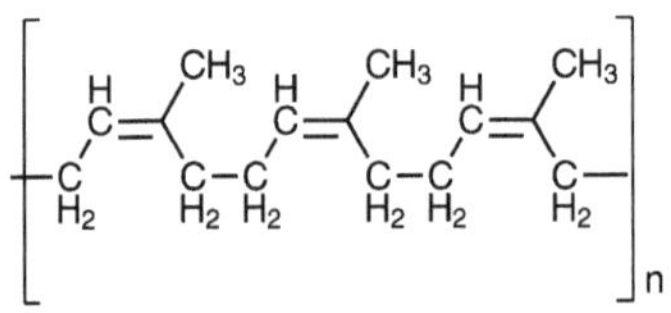

Figure 1.1: Molecular structure of NR.

Although NR is grown in many countries nowadays, at the time of writing the main production areas are in South East Asia, principally Thailand, Indonesia and Malaysia, Vietnam and India. There is also significant production in China, Sri Lanka, the Philippines and Cambodia, with smaller amounts in Africa (principally Liberia) and Central America. The production of NR latex in its original home in the Brazilian rainforests is still low, in part because of a fungal infection, South American leaf blight, to which the rubber tree is prone.

NR latex is present in the tree in specialised cells known as latex vessels, located just behind the outer bark of the tree. Synthesis of the polymer by the tree follows a complicated biological pathway and is not the result of isoprene monomer polymerisation. Instead, naturally occurring carbohydrates are converted *via* a complicated biological process into isopentenyl pyrophosphate, an important intermediary in many biosynthetic pathways. Isopentenyl pyrophosphate can, under certain conditions, form polyisoprenes *via* a condensation polymerisation. Detailed descriptions of the biosynthesis of NR in the rubber tree can be found in Natural Rubber Science and Technology by Roberts [2] and Polymer Latices Science and Technology by Blackley [3].

Latex is collected from the tree by the tapping process. A cut is made through the bark into the latex vessels, causing the latex to bleed out into a cup (Figure 1.2). It is important that the cut is not made too deep into the tree, as this would damage the tree and cause the bark to regrow irregularly, making subsequent tapping more difficult. Specialised knives are used to make the cut, and the operation requires a certain amount of skill to carry it out correctly.

Figure 1.2: Tapping the rubber tree.

As the latex vessels spiral clockwise up the tree, the cut will generally be made diagonally downwards, from high on the left-hand side to low on the right-hand side. This way, the maximum number of latex vessels are opened up. The latex flows out readily at first, then slows down and eventually dries up. Tapping is normally carried out early in the morning when the flow of latex is highest and temperatures (and hence the rate of drying) are lower. The yield from each tree depends on many factors, including the particular clone, the size of the tree and the age at which it was first tapped, amongst others. The trees will be tapped again, reopening the cut by shaving off as little of the bark as possible. Although some clones can be retapped daily, retapping generally takes place on alternate days, as the quality of the latex can be reduced by more frequent tapping. The latex will coagulate and dry if left, and is also prone to microbiological spoilage. For this reason, it is usual to put some ammonia into the collecting cup which prevents putrefaction and helps keep the latex liquid. The tapper will return to collect the latex from the collecting cup a few hours later and add it to the latex collected from the other trees. The bulked latex is transferred to the collecting station and from there onto the concentration factory. On arrival at the factory, the latex will be analysed to establish its quality, which will determine whether it is used to produce latex concentrate or dry rubber. Typically, this field latex will have a rubber content of around 30 to 35%. For subsequent use, the latex is concentrated to give a dry rubber content of about 60 to 65% – above this the viscosity rises rapidly and the latex becomes increasingly difficult to handle and process. The rubber particles have a density of around 0.92 kg/m^3 and so are lighter than the aqueous phase in which they are dispersed. This density difference allows the latex to be concentrated by centrifuging. Latex can also be concentrated by creaming or evaporation, although most latex used for dipping is centrifuged. Centrifuged NR latex is usually supplied with one of two preservative systems: the so-called 'high ammonia' (HA) latex (also known as 'full ammonia') with a minimum ammonia level of 0.60% m/m, and 'low ammonia' (LA) latex with a maximum ammonia content of 0.29% m/m. This lower ammonia content is too low to give adequate protection against microbial spoilage, and a secondary preservative system is employed with LA latex – either a mixture of tetramethylthiuram disulfide and zinc oxide (ZnO), sodium pentachlorphenate or boric acid. These LA latices will be used for applications where the smell of ammonia must be minimised. For some applications, the centrifuged latex will be diluted back to the original solids content of around 30% and centrifuged again. Such latex is, not surprisingly, known as 'double-centrifuged' (DC) latex. A partial dilution of either creamed or centrifuged latex to about 45% total solids, followed by a second centrifugation step can also be used, and latex from this process is known as 'substage' latex. The specification for the latex will be agreed between the vendor and the customer, or the requirements for centrifuged latex in the International Organization for Standardization (ISO) 2004:2010 [4] as shown in Table 1.1, can be applied.

As well as the rubber polymer, NR latex contains many other materials collectively known as 'non-rubber materials'. The composition of field latex varies considerably, depending on the clone, soil, season and so on, but a typical composition of field latex is given in Table 1.2 and compared with a centrifuged HA latex.

The effect of repeated concentration and dilution is to reduce the level of these non-rubbers in the latex, and both DC and substage latex will have lower proportions of proteins and resins and so on in the latex, which can affect the way the

Table 1.1: ISO 2004 requirements for NR latex concentrate.

Property	Test method	Centrifuged latex	
		HA latex	LA latex
Total solids content	ISO 124	61.5% min.	61.5% min.
Dry rubber content	ISO 126	60.0% min.	60.0% min.
Non-rubber solids	–	2.00% max.	2.00% max.
Alkalinity (as NH3)	ISO 125	0.6% min.	0.29% max.
MST	ISO 35	650 s. min.	650 s. min.
Coagulum content	ISO 706	0.05% max.	0.05% max.
Copper content	ISO 8053	8 ppm max.	8 ppm max.
Manganese content	ISO 7780	8 ppm max.	8 ppm max.
Sludge content	ISO 2005	0.10% max.	0.10% max.
Volatile fatty acid number	ISO 506	0.20 max.	0.20 max.
KOH number	ISO 127	1.0 max.	1.0 max.
Colour	Visual inspection	No pronounced blue or grey	
Odour (after neutralisation)	–	No pronounced odour of putrefaction	

MST: Mechanical stability time
Reproduced with permission from ISO 2004:2010 Natural rubber latex concentrate – centrifuged or creamed, ammonia-preserved types – Specifications, International Organization for Standardization, Geneva, Switzerland [4].
Permission to reproduce extracts from British Standards is granted by BSI Standards Limited (BSI). No other use of this material is permitted. British Standards can be obtained in PDF or hard copy formats from the BSI online shop: www.bsigroup.com/Shop.

Table 1.2: Typical composition of field latex and latex concentrate.

	Field latex	HA latex
Rubber	30–35%	60%
Proteins and so on*	1–2.5%	1%
Resins and soaps	1–2.5%	0.4%
Salts	Generally low	0.4%
Ash	<1%	0
Ammonia	0	0.7%
Water	To 100%	To 100%

*Proteins, sugars, amino acids and other carbohydrates

latex behaves in the factory. It has been found that vulcanisation and prevulcanisation proceed more rapidly in the presence of some of the naturally occurring proteins and amines [5], and DC latex is less stable than single-centrifuged latex, presumably due to the loss of some natural stabilisers. Some companies use substage or DC latex in the belief that the latex is more 'pure' and better suited for use in medical devices, such as condoms and surgical gloves. However, if such formulations require additional surfactant and accelerator to compensate for these changes in reactivity and stability then any gains in 'purity' are questionable. It is recommended that any company wishing to use substage or DC latex carries out trials to confirm that any improvement in properties is worth the extra material cost.

1.1.2 Styrene-butadiene rubber latex

Butadiene (specifically 1,3-butadiene, Figure 1.3) and styrene (Figure 1.4) can be copolymerised through an emulsion polymerisation process to give SBR latices. The ratio of the two monomers can be varied, and those with a high proportion of butadiene give more rubbery polymers, whilst those with a high proportion of styrene yield more rigid glassy polymers. Thus, styrene-butadiene copolymers cover a wide range of materials with an extensive range of properties. Whilst SBR latices can form a coherent film, especially those with a higher butadiene content, the level of branching and crosslinking present in commercially available materials mean that these copolymers do not have the stereoregular structure of NR, and therefore cannot crystallise on stretching. As crystallisation can lead to significant increases in strength, the absence of such crystallisation means that the physical strength of these SBR elastomers is very much lower than NR, and will require compounding with reinforcing fillers for any application where strength is important. Effective reinforcement of polymers in latex form is difficult and for this reason SBR latices are rarely used in dipping applications. SBR latices, however, can have some value as reinforcing fillers in other latices, as discussed further in Chapter 5.

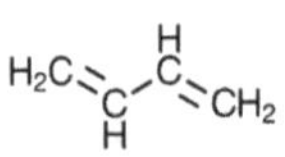

Figure 1.3: Molecular structure of butadiene.

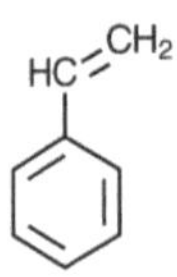

Figure 1.4: Molecular structure of styrene.

1.1.3 Acryonitrile-butadiene rubber latex

NBR latex, also known as 'nitrile' latex, is the major synthetic latex used in the production of dipped articles. Many gloves used in the industrial and food handling sectors are currently made from NBR latex, as are many surgical and examination gloves. One of the key attributes of the material, compared with NR, is that the material is inherently free of the NR proteins, which may cause a reaction in people who are sensitised. As with SBR latex, butadiene can be copolymerised in an emulsion with another monomer – in this case acrylonitrile (Figure 1.5) – to give a latex elastomer.

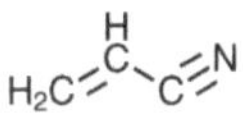

Figure 1.5: Molecular structure of acrylonitrile.

However, the physical properties of NBR elastomers are significantly better than those of SBR. In particular, the presence of polar acrylonitrile molecules in the polymer backbone gives the elastomer much better resistance against non-polar materials such as oils, and for this reason NBR latex is the material of choice for many industrial glove applications. The higher the proportion of acrylonitrile in the copolymer the more like a plastic the material feels (and the higher the oil resistance), and conversely a greater proportion of butadiene makes the material more rubber like. Typically, the acrylonitrile content in NBR latex for glove use will be in the range of approximately 18 to 45%. Latices containing around 18 to 25% of acrylonitrile are often referred to as 'low acrylonitrile' latices, 26 to 40% are called 'medium acrylonitrile' and above 40% 'high acrylonitrile', although these figures are arbitrary. Films dipped from NBR latex have a lower degree of surface tack than those prepared from NR latex, which means that only gloves dipped from those latices with lower acrylonitrile content can be easily beaded. In addition, the films deposited from NBR latex tend to have a low wet-gel strength, and problems with cracking and splitting can occur if the dipping and drying conditions are not closely controlled. Many of the NBR latices used for dipping are carboxylated, i.e., a carboxyl group is introduced into the polymer backbone, for example, by including a small amount of methacrylic acid (MAA) (Figure 1.6) into the formulation.

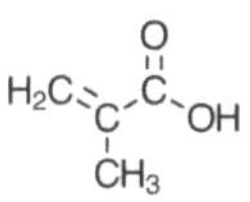

Figure 1.6: Molecular structure of MAA.

Having these ionic groups in the polymer backbone allows crosslinking of the polymer *via* ionic mechanisms using ZnO or an alkali such as potassium hydroxide (KOH) [6], although the butadiene also allows crosslinking through reaction with sulfur, as in NR vulcanisation. Crosslinking with materials such as melamine formaldehyde resins and urea formaldehyde resins has also been reported [7]. Ionic crosslinking can lead to an elastomer with higher tensile strength than sulfur vulcanisation, although ionic bonds can creep under stress, resulting in a higher tensile set. It is not uncommon for an NBR latex formulation to use both sulfur and ionic crosslinking to take advantage of the benefits conferred by each system. Although NBR elastomers can have a very high tensile strength, the tear strength is much lower than other elastomers such as NR or CR, which can be important in many applications. The material also typically has a higher modulus than NR, and one consequence of this in dipping can be considerable stresses build up in the rubber film as it dries, water is lost and the film tries to shrink. Owing to the low wet-gel strength, this can lead to film cracking and, if the former carries a pattern, double imaging. As mentioned above, the polar nature of the NBR backbone means that films of this material exhibit good resistance to oils and greases, but are less resistant to polar materials such as acids, alkalis or ketones.

1.1.4 Polychloroprene latex

CR (often known as 'Neoprene', after the trade name of the original DuPont material) is another synthetic elastomer widely used in latex dipping. It is synthesised by the emulsion polymerisation of chloroprene (2-chloro-1,3-butadiene) (Figure 1.7).

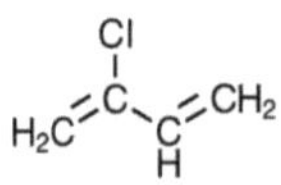

Figure 1.7: Molecular structure of chloroprene.

The presence of chlorine atoms on the polymer backbone means that the polymer is more polar than NR, and hence has a better resistance to oils, but is not as resistant as most NBR latices. CR latices can gradually liberate hydrochloric acid during storage, as the carbon–chlorine bonds in the polymer are hydrolysed. This in turn leads to the pH of the latex gradually falling, which can destabilise the latex. For this reason CR latices usually have a high pH – typically 12 or higher when fresh – to offset the effect of any release of hydrogen chloride, which is called the 'alkali reserve'. The presence of these labile chlorine atoms on the polymer means that CR latices can be crosslinked by metal oxides, usually ZnO. However, the presence of unsaturation in the polymer chain means that sulfur vulcanisation is also possible. As with NBR, sulfur, or a sulfur donor, together with an organic accelerator, is often used in the formulation to optimise the properties. The structure of the chloroprene monomer is such that it can be incorporated into the polymer chain in several different ways during the polymerisation reaction, as shown in Figure 1.8.

The consequence of this is that the finished polymer can contain significant amounts of branching and crosslinking. The effect of this microstructure is to

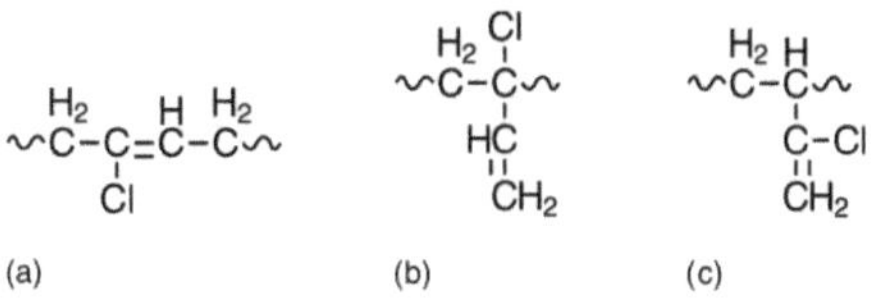

Figure 1.8: The different ways in which the chloroprene molecule can be incorporated into the polymer backbone a) 1,4-addition, b) 1,2-addition and c) 3,4-addition.

interfere with the ability of the polymer backbone to crystallise, which in turn will affect the physical properties of articles formed from the polymer. CR rubbers have a tendency to crystallise readily, especially at low temperatures, and this tendency can be modified by controlling the polymerisation conditions to change the way in which the chloroprene monomer is incorporated into the polymer chain. Polymers showing a low tendency to crystallise will typically have a lower modulus and tensile strength than those polymers that can crystallise more easily. Dipped CR articles will generally have a higher modulus than those dipped from NR, together with better ozone and oil resistance (although the oil resistance is inferior to that of most NBR compounds).

1.1.5 Artificial latices

As well as NBR and CR, there are other synthetic elastomers used in dipping. However, these polymers cannot be synthesised through an emulsion polymerisation route, but rather through solution polymerisation. If they are to be dipped as latices, to take advantage of a latex system, i.e., high solids content and a low viscosity, these polymer solutions need to be converted to a latex. There are several ways in which this can be accomplished, but it usually involves preparing a solution of the polymer, which is then emulsified and the original solvent removed by steam distillation or evaporation under a vacuum. A consequence of this form of preparation is that there can be a higher level of surfactant in the latex, as the surfactant used to emulsify the original solvent will remain after the solvent has been removed. This can make artificial latices very stable, and can also impart a degree of water sensitivity to the finished film. Nevertheless, there are some polymers whose properties are such that it is worth going to the trouble and cost of this preparation method, although the only material with a significant usage in the dipping industry at present is IR.

1.1.5.1 Synthetic *cis*-1,4-polyisoprene latex

The synthetic form of NR, produced by the polymerisation of the isoprene monomer (Figure 1.9), has received much more interest in recent years as it – theoretically at

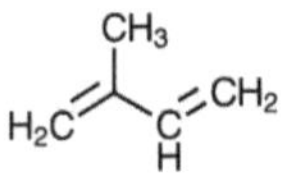

Figure 1.9: Molecular structure of isoprene.

least – can offer the advantages of NR without the risk of allergy to the associated latex proteins. It is now possible to buy high-quality IR latex and it is being increasingly used to produce protein-free gloves and condoms. The main drawback with IR elastomers produced using free-radical polymerisation was that the resultant polymer did not have anything like the stereoregular structure of NR, inhibiting the possibility of crystallisation and hence the physical properties were much worse than NR.

Recent advances in polymer synthesis have meant that isoprene can be polymerised by stereospecific procedures such as Ziegler–Natta polymerisation using a titanium/ aluminium catalyst, or anionic polymerisation initiated by an alkyl lithium initiator. Both of these processes can give IR with a high proportion of *cis*-1,4 content, although not as high as the approximately 99% of NR. As IR latex does not contain any of the proteins and other non-rubber constituents of NR it can give a very clear film, especially the polymer prepared *via* anionic polymerisation. However, the absence of these materials has other effects, not least on physical properties. In general, compared with the corresponding NR film, the modulus, and tensile and tear strengths of IR films are lower whereas the elongation is higher. This is attributed to the reinforcing effect that proteins can have on the material, and the small amount of *trans*-isomer inhibiting crystallisation to some extent. The non-rubber materials in NR latex can also help accelerate the prevulcanisation reactions, and therefore it is no surprise to find that IR latex is more difficult to prevulcanise than NR latex and requires careful formulation and control of the rate and extent of any prevulcanisation. However, despite these problems, and the high cost of the material, IR latex is being increasingly used to produce good quality gloves and condoms that are free of extractable latex proteins.

1.1.5.2 Isoprene-isobutylene rubber (or butyl rubber) latex

IIR is a copolymer of isobutene copolymerised with a small amount of isoprene to give sufficient unsaturation to allow it to be vulcanised using conventional sulfur vulcanisation technology (Figure 1.10).

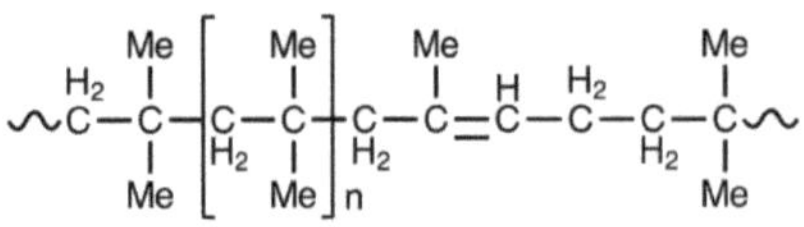

Figure 1.10: Molecular structure of butyl rubber.

IIR has a very low permeability to gases, and as such is of interest as a material for protective gloves, especially against chemical and biological warfare materials. It also has a good resistance to ozone. However, its very low gas permeability means that drying a dipped film can be difficult and it will usually be necessary to build up the film thickness with multiple thin dips. It is commonly used at present as a speciality coating onto fabrics to improve barrier properties.

1.1.5.3 Chlorosulfonated polyethylene latex

CSR latex (also known as Hypalon from the trade name of the DuPont material) is another artificial latex developed to offer specific properties, in this case excellent chemical and ageing resistance. Its structure is a polyethylene chain with chlorine atoms and SO_2Cl groups attached to the polymer backbone at intervals (Figure 1.11).

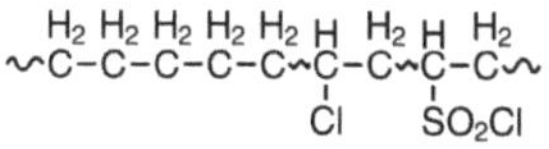

Figure 1.11: Molecular structure of CSR.

The amount of chlorine and sulfur in the polymer depends on the grade of material, but can range from 25 to 45% for the chlorine content and from 1 to 1.5% for the sulfur content. Like CR, it can be crosslinked *via* the chlorine atoms by means of metal oxides such as magnesium oxide or ZnO. Like the artificial IIR latex, its main use is as a speciality coating material.

1.1.5.4 Polyurethane

PU are a very wide category of polymers, whose only criterion is that there is a urethane linkage (Figure 1.12) in the polymer chain.

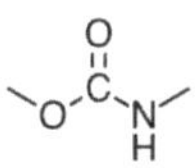

Figure 1.12: Urethane linkage.

By modifying the other components of the polymer backbone, those linked by the urethane chain, PU elastomers can be made which display a wide variety of physical properties to meet a range of end uses. By incorporating specific chemical groupings into the polymer chain, stable PU latices can be prepared without the need for surface active agents. However, despite the obvious advantages offered by PU latices, they do not seem to be used in the latex dipping industry.

1.1.5.5 Ethylene-propylene rubber latex

Ethylene-propylene rubber latex latices are artificial latices which, like CSR latices, offer a high level of chemical resistance, owing to the inert nature of the polymer chain. To enable the material to be crosslinked, a small amount (up to approximately 10%) of a diene monomer is copolymerised into the polymer chain to give EPDM. Although EPDM latices were available in the past, at the time of writing, there are not thought to be any commercial supplies available.

1.1.6 Other naturally occurring latices

The *Hevea* tree is not the only source of *cis*-1,4-polyisoprene latex, and in recent years there has been considerable work focused on exploiting other sources. Chief amongst these are latices from the guayule shrub *Parthenium argentatum*. Guayule grows in the arid regions of the Southern US and Northern Mexico, and in other similar areas. Unlike the rubber tree, the latex is present in cells within the bark, not in special latex vessels. In order to extract the latex the plant has to be crushed, and the latex separated, purified and concentrated. Whilst originally the whole plant was macerated, now only the branches are harvested, leaving the stem to regrow. Thus it is no longer necessary for a new plant to grow, and higher yields can be obtained. Unlike the *Hevea* tree, where the tapping and latex collection is labour intensive, latex collection from the guayule plant can be mechanised. The main attraction of *cis*-1,4-polyisoprene from guayule is that it contains a lower level of naturally occurring proteins than *Hevea* rubber, and the proteins that are present do not appear to cause any allergenic reactions in humans. It therefore offers a route to produce hypoallergenic NR latex for gloves and condoms. At the time of writing, guayule latex is in commercial production.

Another source of NR is the Russian dandelion, *Taraxacum kok-saghyz*. In the case of this plant, the latex is contained in a large tap root. Commercial production of NR latex from this plant is less advanced than that from the guayule plant, and at the time of writing, it is not thought that there is any commercial supply.

1.2 Alkalis

The pH of a latex is important, as this helps control stability through electrostatic repulsion and also helps maintain surfactants such as fatty acid soaps as salts, rather than as the free acid. The most commonly used alkalis are KOH and ammonium hydroxide (NH_4OH) (although sodium hydroxide is usually used with CR latices), and these differ in one important respect. NH_4OH is volatile and will be lost from the latex over a period of time, whereas KOH is not volatile and its effects in the latex are permanent. Consequently, KOH is often referred to as 'fixed alkali' and NH_4OH as 'fugitive alkali'. NH_4OH can also affect latex stability – in particular the stability of NR latex – through the zinc ammine reaction, described in more detail in Chapter 6, which can be important in heat-sensitised dipping. This zinc ammine reaction, and the associated ZnO thickening, can be reduced by the presence of KOH. In straight dipping, for example, the production of NR condoms and examination gloves, ammonia will be lost fairly quickly from the thin film immediately after dipping, especially in the presence of heat. Loss of ammonia can speed up the rate of increase in viscosity of the latex and help it to 'set up' on the former. This in turn will restrict flow and help promote an even film thickness. Hence, a well-designed latex compound will usually be formulated with a mixture of NH_4OH and KOH to give the correct balance of stability and speed of drying.

The ideal pH range for the latex will depend upon the latex and the dipping conditions, but will generally be in the range of 10.0 to 10.5, although as mentioned above, CR latices will have a pH of 12 or higher to neutralise any hydrochloric acid released from the polymer backbone. However, it is worth noting that the viscosity of carboxylated latices such as some XNBR latices can increase markedly if too much alkali, especially KOH, is added.

Finally, NH_4OH can be invaluable as a stripping aid for NR products. If the products are difficult to strip, for example, if the latex is fresh or the drying temperatures are too high, then a small amount of ammonia added to the leach tank will ease the stripping to a great extent. However, if ammonia is to be used in this way, there must be adequate extraction on and around the dipping line to reduce the level of ammonia fumes in the atmosphere.

1.3 Surfactants

Surfactants, or surface active agents, are generally essential in a latex formulation, although for some applications and formulations the inherent stability of the latex is sufficient. Their job is to help balance the stability of the latex so that it can be successfully dipped. If the latex is not stable enough then it can start to coagulate during handling or in the dip tank. If the latex is too stable it will not form a film quickly enough on the former, leading to excessive flow, uneven film formation and film cracking. The mechanisms that are brought into play by the different surface active materials will be covered in more detail in Chapter 6. There is a bewildering array of surface active agents available today, but fortunately there are relatively few classes of materials widely used in the latex industry.

First of all there are fatty acid soaps, which are formed by the reaction of a fatty acid with an alkali (Figure 1.13). The potassium soap is the most widely used, as potassium salts are usually more soluble than those with other cations. The most commonly used fatty acid soaps have chain lengths ranging from 6 (caproic) to 12

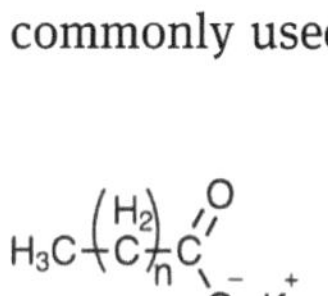

Figure 1.13: General structure of a saturated fatty acid soap.

Table 1.3: Chain lengths of fatty acids.

Trivial name	Carbon chain length	Saturated
Caproic acid	6	Yes
Caprylic acid	8	Yes
Capric acid	10	Yes
Lauric acid	12	Yes
Myristic acid	14	Yes
Palmitic acid	16	Yes
Oleic acid	18	No – 1 double bond
Stearic acid	18	Yes

(laurate) (Table 1.3). Generally, the fatty acid backbone is saturated, although unsaturated soaps such as oleates are also used.

An important property of fatty acid soaps when used to stabilise latex formulations is their ability to improve stability without interfering with gelation of the latex. Synthetic alternatives to natural fatty acid soaps are available, which are claimed to give superior performance in latex dipping [8]. Fatty acid soaps will react with multivalent metal ions, and this property is utilised in coagulant dipping where the former is coated with (usually) a calcium compound. The calcium ions react with any fatty acid soaps, converting them into the calcium salt and rendering them insoluble, destroying their stabilising effect. A similar reaction occurs, less helpfully, in NR latices containing ammonia and a zinc-containing material such as ZnO or a zinc-based accelerator. In these cases the zinc ions may react with the fatty acid soaps through the zinc ammine reaction (Chapter 6) and cause the latex to increase in viscosity – the so-called 'ZnO thickening'. Another unwanted property of fatty acid soaps, especially those with a longer chain length, is that they can stabilise the liquid–air interface and act as foam promoters. This will help the inclusion of stable air bubbles in the latex and also lead to the generation of foam at, for example, the stripping station where any residual surfactant can be extracted from the rubber film. Both ZnO sensitivity and the extent to which they enhance stability also depend on the chain length of the fatty acid soap. Those with a shorter chain length, for example, C8 (caprylate) are less prone to ZnO thickening, whereas fatty acid soaps with longer alkane chains, say C12 and above, form insoluble zinc ammine salts, and therefore can induce stability problems in the presence of ZnO.

There are many other anionic surfactants available, such as the alky aryl sulfonates, although these tend to be used more as wetting agents in the preparation of dispersions, the common example being salts of the condensation product of naphthalene sulfonic acid and formaldehyde.

Non-ionic surfactants are also encountered in latex dipping, both in the preparation of emulsions and as a stabiliser in the latex formulation itself. The most widely used are usually ethylene oxide (C_2H_4O) condensate materials, especially those with longer C_2H_4O chains. These are very powerful and effective materials for stabilising latices, but should be used with caution in dipping formulations as they can promote excessive flow and impair gelation of the latex if used in too high a concentration. In general, they are not sensitive to pH, nor to zinc or calcium ions. The optimum concentration is frequently very low, and careful experimentation is recommended when incorporating them into a formulation for the first time. A particular use of a non-ionic stabiliser is in heat-sensitised dipping, where the negative solubility parameters of some ethoxylates (which become less soluble in water as the temperature is increased) is used to give a controlled destabilisation of the latex as the temperature is raised.

Latex compounds can also be stabilised by water-soluble hydrocolloids, such as casein, alginates or cellulose derivatives. These different types of stabiliser can bring into play different stabilisation mechanisms, as discussed in Chapter 6.

The range of surface active materials is extremely wide and this section has only touched on the ones more commonly encountered in latex dipping. For a more specialised treatment the reader is directed towards one of the many reference books on the subject, or to the larger suppliers who maintain a technical assistance department.

Finally, some of the synthetic latices are carboxylated and these carboxylic acid groups can play a major part in stabilising the polymer chain. This reduces the need for added surfactants to stabilise the latex, but can result in the finished article having some degree of water sensitivity.

1.4 Accelerators

Accelerators for latices are generally much more active than those used for dry rubber. In dry rubber the processing temperatures encountered in, for example, an internal mixer or two-roll mill, are much higher than can be used for latices, and consequently the more active accelerators would rapidly lead to the compound scorching. Latex technology uses much lower temperatures and some of the accelerators are active at room temperature when correctly formulated. Consequently, the range of accelerators encountered in latex formulations is fairly limited. Dithiocarbamates are the accelerators used most often, with thiurams, thiazoles, sulfenamides and thioureas also being used on occasion.

1.4.1 Health and safety of accelerators

The majority of accelerators used in the latex industry are the product of a reaction between carbon disulfide and an amine, and many of them do not have a good safety profile. Rubber accelerators tend to be consumed during the manufacture of dipped latex products, and any accelerator residues in the finished article will usually be very low and located within the rubber film. A greater potential risk is when the accelerators are handled in the factory, and they should be treated with care and knowledge of any specific risks as listed in the relevant material safety data sheet (MSDS). It is recommended that personnel handling rubber accelerators use adequate personal protective equipment, for example, dust masks, gloves and eye protection, and that there is adequate dust control and extraction in the storage and compounding areas, and other areas where exposure to the accelerators may occur.

1.4.2 Types of accelerators

1.4.2.1 Dithiocarbamates

As noted above, dithiocarbamates are by far the most widely used accelerators for latices, especially NR latices. They are fast accelerators, which can give a good balance of physical properties in both natural and synthetic latices. They can be used as a dispersion of the zinc salt or a solution of the sodium, or more rarely, ammonium salt. Dithiocarbamates have the general structure shown in Figure 1.14.

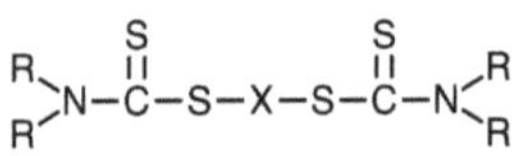

Figure 1.14: General structure of dithiocarbamate accelerators. 'X' represents a divalent metal atom and 'R' an alkyl or aryl grouping.

The most commonly used dithiocarbamates are the diethyl and di-*n*-butyl dithiocarbamates, usually as the zinc or sodium salts (Figures 1.15 and 1.16, respectively), although dithiocarbamates with a large alkyl group, such as the diisononyldithiocarbamate (Figure 1.17) are used for specific products such as those where very low extractable *N*-nitrosamines are required.

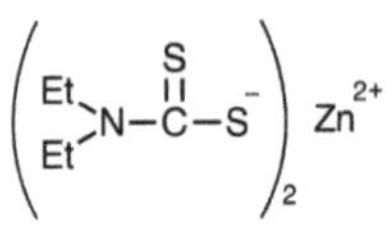

Figure 1.15: Zinc diethyldithiocarbamate (ZDEC).

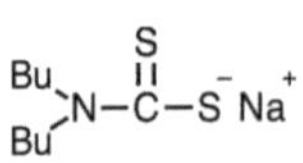

Figure 1.16: Sodium di-*n*-butyldithiocarbamate.

Similarly, aromatic groupings, such as dibenzyl (Figure 1.18), can be used to produce dithiocarbamates that generate low extractable *N*-nitrosamines. In common

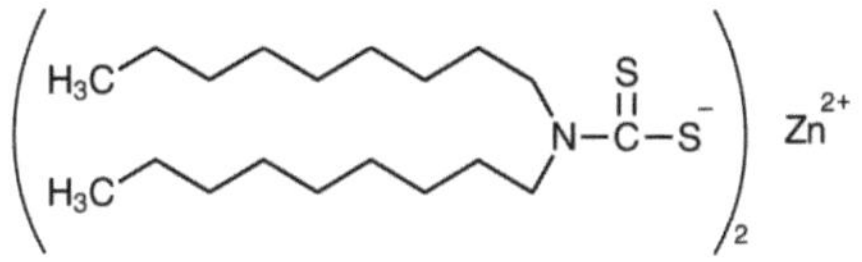

Figure 1.17: Zinc diisononyldithiocarbamate.

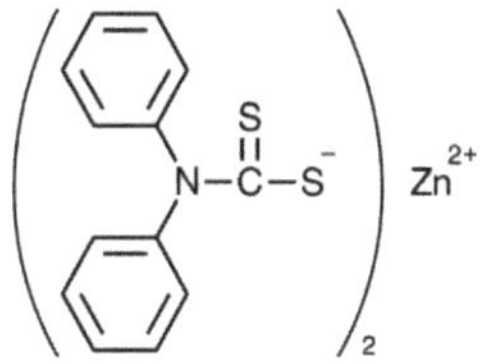

Figure 1.18: Zinc dibenzyldithiocarbamate.

with most rubber accelerators, dithiocarbamates have the potential to cause type IV allergic contact dermatitis in contact with skin, and it has been found that in general, the higher the MW, the less likely the material is to provoke an allergic reaction. For example, a study comparing the relative health hazards of zinc dimethyl, diethyl and dibutyldithiocarbamates in gloves concluded that the accelerator residues from the use of zinc di-*n*-butyldithiocarbamate (ZDBC) had the least potential to cause harm [9]. However, zinc dimethyldithiocarbamate has a higher specific gravity than the other dithiocarbamates (1.66, compared with 1.47 for ZDEC and 1.27 for ZDBC), which allows it to be more easily removed from a compounded latex by centrifugation or sedimentation, thus reducing the accelerator concentration in, and hence the allergenic potential of, the finished product. It is also important to note that residual dithiocarbamate accelerators are effective antioxidants, so removing them from the product may affect its environmental stability. As in so many instances, latex technology requires compromises to be made. Dithiocarbamates are fast accelerators, although their activity depends on the alkyl or aryl grouping on the dithiocarbamate moiety [5]. Figure 1.19 shows the effect of the alkyl chain length of zinc di-*n*-alkyldithiocarbamates on the reduction of free sulfur in NR latex at 70 °C. The numbers beside each curve show the number of carbon atoms in the alkyl grouping. It can be seen that the reduction of free sulfur with zinc dimethyldithiocarbamate ($n = 1$) is relatively slow, the diethyl salt ($n = 2$) is much quicker and the dibutyl salt ($n = 4$) is the fastest of all. After that the reaction rate slows down with increasing chain length, which is believed to be the result of two conflicting factors. As the chain length increases, the solubility of the dithiocarbamate (or the activating species derived from it) in the liquid phase, and hence the ease with which it is transported to the rubber particle, decreases; however, the compatibility of the accelerator with the rubber particle increases. ZDBC thus shows the best balance between these factors.

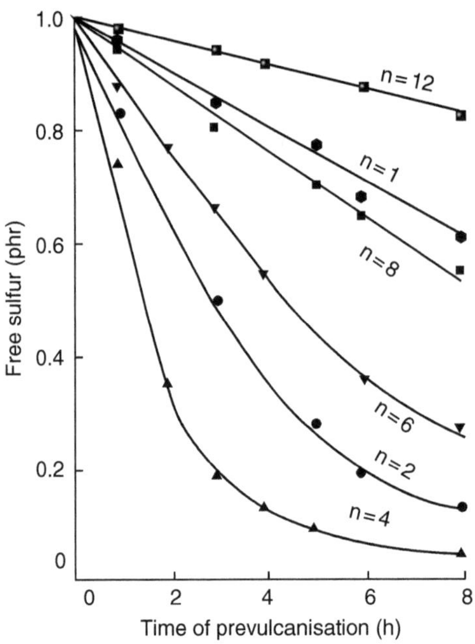

Figure 1.19: The effect of dithiocarbamate chain length on rate of prevulcanisation. Phr: parts per hundred of rubber. Adapted from A. C-P. Loh in *Further Investigations of the Prevulcanization of Natural Rubber Latex*, The North London Polytechnic, London, UK, 1982 [PhD Thesis] [5].

For this reason, together with the reduced risk of adverse health reactions mentioned earlier, ZDBC is one of the most commonly used accelerators in latex systems.

The efficiency of the dithiocarbamate accelerator can be enhanced by the use of activators, or synergistic combinations of accelerators, which can allow a lower accelerator level to be used to obtain the required balance of physical properties. Common synergistic couplings with dithiocarbamates are thiazoles or amines, and proprietary activated dithiocarbamates are available from several suppliers.

1.4.2.2 Thiuram disulfides and polysulfides

Thiuram disulfides have the general structure shown in Figure 1.20.

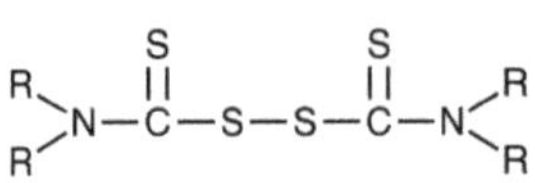

Figure 1.20: General structure of thiuram disulfides.

As can be seen, the structure is similar to the dithiocarbamates, and indeed the vulcanisation pathway is thought to be the same. Thiuram disulfides and polysulfides (Figure 1.21) can be used to vulcanise unsaturated rubbers without using elemental

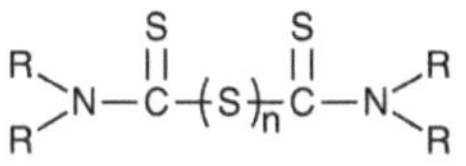

Figure 1.21: General structure of thiuram polysulfides.

sulfur, although the vulcanisation reaction is slow when compared with that using dibutyl or diethyldithiocarbamates in conjunction with sulfur. The vulcanisation reaction can, however, be activated by certain sulfur compounds such as thiourea and *N,N′*-diethylthiourea (Figure 1.22). A combination of 1:1 (mole ratio) of thiourea and tetramethylthiuram disulfide is claimed to result in fast vulcanisation at temperatures below 100 °C [10]. Vulcanisates from the thiuram di- and polysulfides without elemental sulfur show improved resistance to heat and oxidation, especially oxidation catalysed by copper.

Figure 1.22: Thiourea and *N,N′*-diethylthiourea.

In general, however, thiuram sulfides are used in latex vulcanisation as a secondary accelerator with a dithiocarbamate and elemental sulfur.

1.4.2.3 Thiazoles

Thiazoles, especially zinc 2-mercaptobenzothiazole (ZMBT) (Figure 1.23) and the sodium salt, sodium 2-mercaptobenzothiazole (SMBT) (Figure 1.24) are sometimes encountered in latex formulations, but almost invariably as secondary accelerators in combination with dithiocarbamates. The combination of these two accelerators, usually in the proportions between 1:9 to 1:3 thiazole to dithiocarbamate, show considerable synergism, resulting in faster vulcanisation and, especially in the case of NR, a significant increase in modulus. As with other accelerators, the sodium salt is soluble in water whilst the insoluble zinc salt must be added in the form of a dispersion. It is possible to use 2-mercaptobenzothiazole (MBT) itself (Figure 1.25), together with ZnO, as the secondary accelerator, but the acidic thiol group will tend to destabilise the

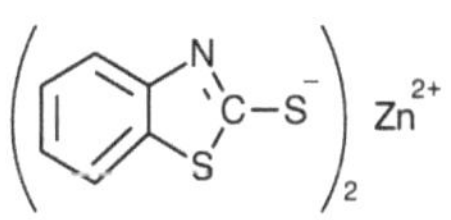

Figure 1.23: ZMBT.

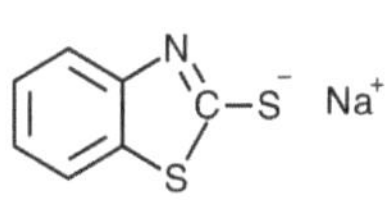

Figure 1.24: SMBT.

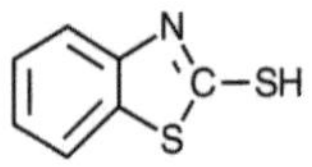

Figure 1.25: MBT.

latex, quickly causing thickening or even gelation. If these materials are to be employed as secondary accelerators, it is strongly recommended that the zinc or sodium salt is used, and even these can cause stability problems if not properly formulated. One of the characteristics of thiazoles is that, as the nitrogen is present as a tertiary amine, it will not give rise to *N*-nitrosamines, which are formed from secondary amines such as dithiocarbamates and thiuram disulfides. It can therefore have applications in building a formulation with a reduced tendency to form *N*-nitrosamines. Note, however, that MBT and its derivatives also have a very bitter taste, limiting their use in products such as baby bottle teats and dental gloves, or any other article likely to come into contact with the mouth. Both MBT and ZMBT have been placed on a list of chemicals that 'probably cause cancer', hence their use in consumer articles is not recommended and suitable precautions should be taken whilst handling MBT and derivatives, as outlined in Section 1.4.1 above.

1.4.2.4 Xanthates

Xanthates are another class of accelerators for unsaturated elastomers that will not give rise to *N*-nitrosamines, as they do not contain nitrogen. The structure of three xanthates commonly used for latices are shown in Figures 1.26–1.28: zinc diisopropyl xanthate (ZIX) (Figure 1.26), sodium isopropyl xanthate (SIX) (Figure 1.27) and zinc di-*n*-butyl xanthate (ZNBX) (Figure 1.28). As a class, the xanthates are very rapid accelerators, active even at ambient temperatures; however, they are unstable and readily release the evil-smelling carbon disulfide. In the author's experience, the physical properties – tensile strength, tear strength and modulus – of NR compounds

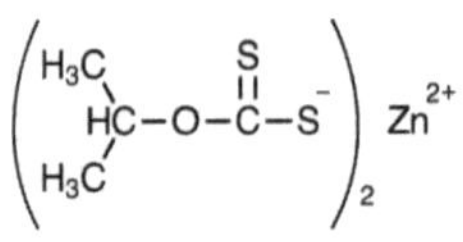

Figure 1.26: ZIX.

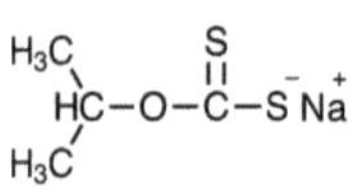

Figure 1.27: SIX.

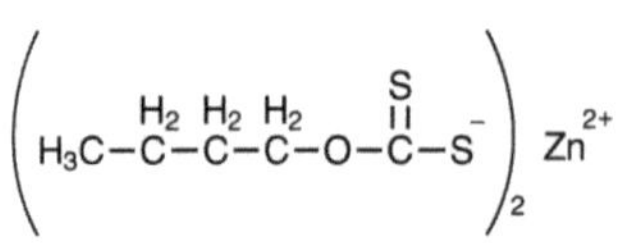

Figure 1.28: ZNBX.

vulcanised with sulfur and ZIX are not as good as compounds made using the more commonly encountered dithiocarbamate/sulfur formulations.

However, there may well be occasions when speed and low vulcanisation temperatures, or the absence of nitrogen, are necessary design considerations, and xanthate-accelerated formulations can prove useful in these situations.

1.4.2.5 Thioureas

N,N'-diphenylthiourea (DPTU) – also known as thiocarbanilide (Figure 1.29) – is one of the earliest rubber accelerators, but is generally too slow to be used as a primary accelerator in rubber latices. However, as mentioned above, it can be used in synergistic combinations with thiuram sulfides and polysulfides, where it has been found to prevulcanise NR latex at temperatures below 100 °C. The main use of DPTU currently is as a coaccelerator with *N,N'*-diphenylguanidine (DPG) for CR latices.

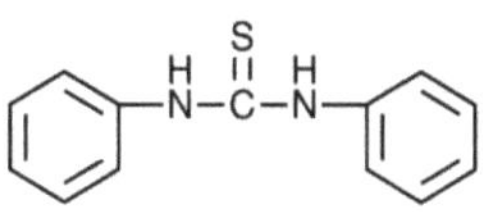

Figure 1.29: DPTU.

1.4.2.6 Guanidines

Guanidines, like thioureas, are some of the earliest accelerators used in the rubber industry, and again are generally too slow to be of interest for latex. However, as mentioned above, DPG (Figure 1.30) is a very effective accelerator for CR latices when used in conjunction with DPTU. Various ratios of the two accelerators can be used, typically from 5: 1 thiourea: guanidine up to 1: 1. The speed of the vulcanisation reaction and properties of the finished products will vary to some extent with the accelerator ratio and concentration, and experimentation is recommended to find the optimum combination for each situation.

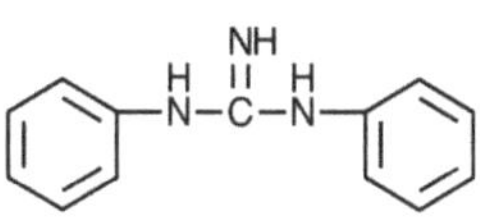

Figure 1.30: DPG.

1.5 Other vulcanisation materials

1.5.1 Sulfur

Virtually all latex elastomers are vulcanised by sulfur. There are some carboxylated or chlorinated elastomers that are crosslinked by metal oxides for specific applications, and occasional uses of peroxides or radiation, but the vast majority of dipped latex articles that are used today owe their strength to sulfur crosslinks (Chapter 3). Sulfur can exist in many different allotropes, but the form most often used in latices is the commonly occurring S_8 ring. Sulfur is one of the more difficult materials to mill into a fine-particle size dispersion, and for that reason a finely divided form known as 'colloidal' sulfur, formed by reacting hydrogen sulfide with sulfur dioxide, is available from some manufacturers. Colloidal sulfur needs little or no milling to be suitable for incorporation into a latex compound, and thus can simplify compounding. It is – naturally – more expensive than the standard material, and it is necessary to note that it is only some 93–95% sulfur, which will need to be taken into account when drawing up the formulation. Both the standard and colloidal forms of sulfur will give equally good products when used correctly. From time to time so-called 'insoluble sulfur' is promoted for use in the latex industry. Insoluble sulfur is a polymeric form of sulfur, formed by rapidly cooling molten sulfur. It is in a metastable state and will revert to the standard S_8 ring with time, especially at elevated temperatures. The author has seen no evidence that insoluble sulfur offers any advantages whatsoever over standard or colloidal sulfur. If the company has the technology to produce an acceptable sulfur dispersion then that will usually be the cheapest option: if not, consider paying extra for the convenience of using colloidal sulfur. Whichever form of sulfur is chosen, it is recommended that good quality sulfur with a high degree of purity is used.

1.5.2 Peroxides

Whilst sulfur crosslinking of elastomers, especially NR, gives vulcanisates that exhibit excellent physical properties, the sulfur crosslinks do not give good high-temperature resistance. However, the carbon–carbon crosslinks that are the result of peroxide vulcanisation impart outstanding temperature resistance. In addition, peroxide-crosslinked materials are free of *N*-nitrosamines and also the accelerator residues that can cause type IV allergic reactions. For this reason there have been many attempts to develop viable vulcanisation and prevulcanisation systems, mainly for NR, based on peroxides [3, 11]. These experiments have shown that there are several significant obstacles to peroxide vulcanisation. The reaction speeds tend to be lower than anticipated, there is frequently the need for a coagent to assist in the reaction and the physical properties are poor, especially at the surface of the article where the peroxide can react with oxygen to promote oxidation of the material. In addition, many antioxidants used in latices function as radical traps and can reduce the effectiveness of the peroxides. Whilst research is continuing to develop a satisfactory peroxide curing system, at the time of writing, it is believed that there are no commercial processes using peroxide curing for dipped latex articles.

1.5.3 Zinc oxide

Whilst ZnO is not essential for vulcanisation, incorporation of ZnO as an activator will markedly improve both the efficiency of the vulcanisation reaction and the physical properties of the finished product. ZnO participates in the vulcanisation reaction by regenerating the active sulfurating agent and helping the crosslinking reactions to proceed. The reaction pathway for the crosslinking of olefinic elastomers is described in more detail in Chapter 3.

In addition to improving vulcanisation, increased levels of ZnO in the formulation can improve resistance to overvulcanisation and will increase the tensile modulus. However, ZnO can also affect the stability of the latex, especially ammonia-preserved NR latex. Adverse reactions to ZnO in these systems will lead to an increase in viscosity (ZnO thickening) and possibly even gelation. This thickening is believed to be caused by the reaction between ammonia and zinc ions, which can in turn react with carboxylic acid stabilisers. Whilst such changes in viscosity can be troublesome, they can also be exploited in, for example, heat-sensitised dipping (Chapter 7) or in straight dipping to help control latex pick-up and film evenness. ZnO thickening is explained in more detail in Chapter 6.

The grades of ZnO commonly used in latex compounding are produced by precipitation and have a small particle size. Fine-particle size dispersions of ZnO are easily prepared by a short time in a ball mill or even by a high-shear mixer. Even finer grades of ZnO – called 'active ZnO' – are offered by some rubber chemical companies. These grades of the material have a very high surface area, which is claimed to enhance their activity. Any such enhancements mean that the required properties could be achieved with a lower loading of ZnO, saving money and improving transparency. As with all claims of this type, the actual improvements achieved on the production line may not match those in the suppliers' advertisements, but for companies looking to reduce the level of ZnO in the product or improve the properties of the finished product, experiments with active ZnO may be worth carrying out.

Away from sulfur vulcanisation ZnO can be used to crosslink polymers containing labile halogen groups or carboxyl groups, such as CR or XNBR latices. This type of ionic crosslinking can give products with a higher modulus and tensile strength than sulfur vulcanisation, although other properties such as tensile set are inferior to sulfur-vulcanised compounds.

1.5.4 Zinc carbonate

The use of ZnO, especially at higher levels, can reduce the transparency of the dipped rubber article. In cases where this loss of transparency is important, zinc carbonate ($ZnCO_3$) may be used as an activator instead of ZnO to give a more transparent film. However, the carbonate is not as active as the oxide, so more may need to be added to achieve the required properties.

The density of $ZnCO_3$, at 4.4, is less than the 5.6 of ZnO and so will not sediment quite as easily in a dispersion, although it will still need agitation to maintain it in dispersion.

1.5.5 Other materials that can activate vulcanisation

Zinc diammine diisocyanate has been reported as functioning as a partial replacement for ZnO in the sulfur vulcanisation of NR latex [12], without the adverse effects on latex stability that can occur with ZnO or $ZnCO_3$. In addition, postvulcanised films prepared using zinc diammine diisocyanate had a higher crosslink density than the control films prepared using ZnO.

The so-called 'multifunctional additives' – amine-bridged amides – have been used to enhance the cure of CR in dry rubber [13], but their applicability to latices has yet to be fully explored.

1.6 Rubber antioxidants

Articles dipped from rubber latices have an inherently better resistance to oxidation than articles moulded from dry rubber, probably because latex is not subjected to mastication on a two-roll mill or in an internal mixer (which breaks down the polymer chain and reduces the MW), nor is it exposed to high processing temperatures. In addition, products such as condoms or surgical gloves are usually packed in impermeable packaging, severely restricting possible exposure to oxygen. Many dipped latex formulations include dithiocarbamate accelerators and residual dithiocarbamate is a powerful antioxidant. Nevertheless, it is a common practice to add an antioxidant to rubber formulations. There are many types of antioxidants available for rubber, some of which are discussed below. In practice, most latex formulations will use hindered phenolic antioxidants in one form or another. Note, however, that some of the hindered phenolic antioxidants are reported to reduce the ozone resistance of CR [14]. The concentration of antioxidant in the formulation will depend upon the product, the application and the antioxidant, but will generally be within the range 0.5 to 2.0 parts per hundred of rubber (phr). Concentrations lower than 0.5 phr are usually ineffective, and unnecessarily high concentrations tend to be discouraged owing to the cost of many antioxidants.

1.6.1 Phenolic antioxidants

One of the requirements of an effective antioxidant is that it is able to capture an oxygen-free radical and delocalise the charge through resonance. Structures that are successfully able to do this are phenolic materials, especially those of higher MW which are less volatile and less extractable, and hence are more persistent in the rubber network. There are numerous phenolic antioxidants available on the market, some of undisclosed composition, but the most commonly used include: antioxidant 2246, antioxidant 425, Wingstay L and butylated hydroxytoluene (BHT).

1.6.1.1 Antioxidant 2246

Antioxidant 2246 – chemically 2,2′-methylene-*bis*-(4-methyl-6-*tert*-butylphenol) (Figure 1.31) – is known by several different names, most of which will contain '2246' somewhere in the name. It is a very effective antioxidant for latices, both natural and synthetic. It is non-staining and relatively non-volatile, and so will persist in the rubber during ageing. The one problem with this antioxidant is that it has a tendency to turn slightly pink with ageing. This 'pinking' can be a problem with pale-coloured articles, although will not be seen if the material is pigmented in any way.

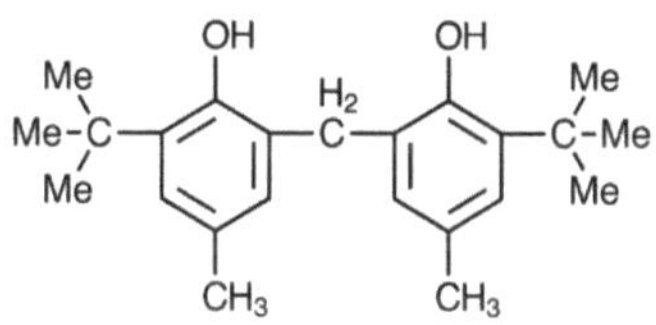

Figure 1.31: Antioxidant 2246.

Antioxidant 2246 is relatively easy to mill to a small particle size, which is not always the case with phenolic antioxidants and it is readily available as a fine-particle size dispersion from several suppliers.

1.6.1.2 Antioxidant 425

Antioxidant 425 (Figure 1.32) is very similar to antioxidant 2245 and exhibits very similar effectiveness as an antioxidant. It also tends to 'pink' with ageing.

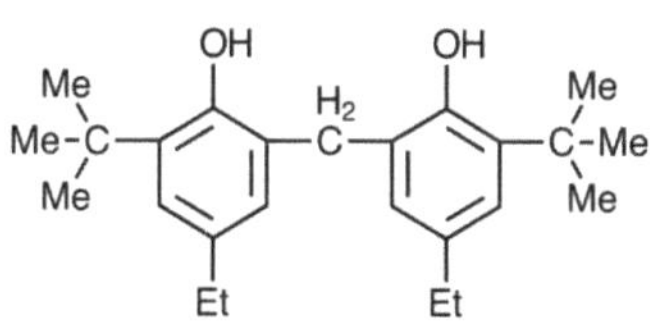

Figure 1.32: Antioxidant 425.

1.6.1.3 Wingstay L

The antioxidant known as Wingstay L, amongst many other proprietary names, is the butylated reaction product of *p*-cresol and dicyclopentadiene. It is a polymeric antioxidant of very low volatility and as such is resistant to extraction. Like antioxidants 2246 and 425, it is very effective in latex systems, especially NR. It is also one of the easier antioxidants to mill to a small particle size and is available as a fine-particle size dispersion from several suppliers.

1.6.1.4 Butylated hydroxytoluene

BHT – chemically 2,6-di-*tert*-butyl-*p*-cresol (Figure 1.33) – is a cheap, readily available and widely used antioxidant for latex systems. Whilst BHT is effective, it is more easily extracted than the other phenolic antioxidants described above, and would be considered less suitable for more demanding applications.

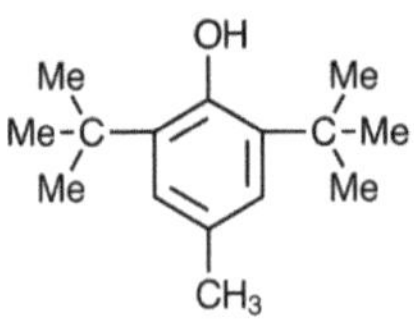

Figure 1.33: BHT.

1.6.2 Liquid antioxidants

There are also some very effective liquid antioxidants available for latex, generally based on styrenated phenols, which have the advantage of not needing to be milled and dispersed before use, simplifying compounding procedures. These are often proprietary materials, sometimes mixtures, and the exact formulation is not usually disclosed.

1.6.3 Para-phenylene diamines

The *para*-phenylene diamines, such as *N,N′*-diphenyl-*p*-phenylenediamine (DPPD) (Figure 1.34) and *N,N′*-di-2-naphthyl-*p*-phenylenediamine (DNPD) (Figure 1.35), are not only very effective antioxidants but also impart some protection against copper and ozone. However, they do discolour during ageing and can stain any light-coloured article in which they are used. They can also cause contact dermatitis, and so, despite their effectiveness, are not often encountered in latex formulations.

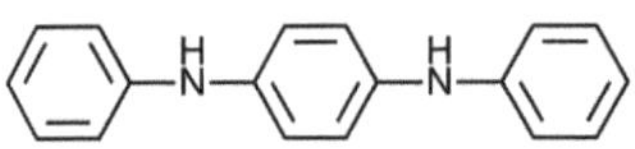

Figure 1.34: DPPD.

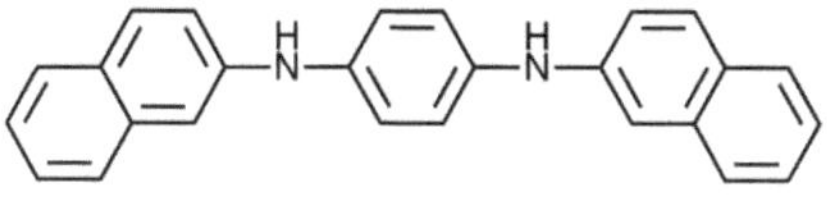

Figure 1.35: DNPD.

1.6.4 Network-bound antioxidants

Loss of the antioxidant during service, either through extraction or volatilisation, can be a problem with some products, especially those in regular contact with water, such as houseware gloves or rubber threads in garments. It is possible to incorporate the antioxidant into the polymer backbone [15–19], elegantly solving all compounding, extraction, migration and volatilisation problems. The type of molecules incorporated into the polymer backbone to confer antioxidant activity are usually *p*-substituted nitrobenzenes, where the substituent is a hydroxyl or amino compounds. Although the claimed performance of these network-bound antioxidants is very good, they do not appear to have been widely used in practice.

1.6.5 Synergistic enhancement of antioxidant activity

It is possible to increase the efficiency of some antioxidants by choosing combinations of antioxidants or combining an antioxidant with another chemical species. One of the most well-known is a combination of a mercaptobenzimidazole compound, for example, mercaptobenzimidazole (MBI) itself (Figure 1.36), or its zinc salt, zinc mercaptobenzimidazole (ZMBI) (Figure 1.37), with amine or phenolic antioxidants [20]. It is possible to buy proprietary dispersions of antioxidant/MBI or ZMBI mixtures. The most common mixture is in the ratio 1: 1 antioxidant: MBI. More recently, zinc 2-mercaptotoluimidazole (ZMTI) (Figure 1.38) has been shown to increase the effectiveness of phenolic antioxidants in NBR [21].

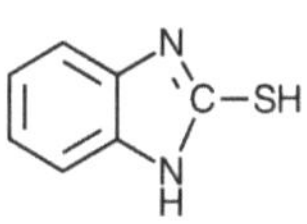

Figure 1.36: MBI.

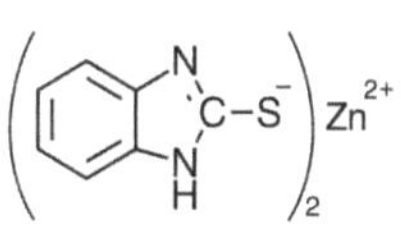

Figure 1.37: ZMBI.

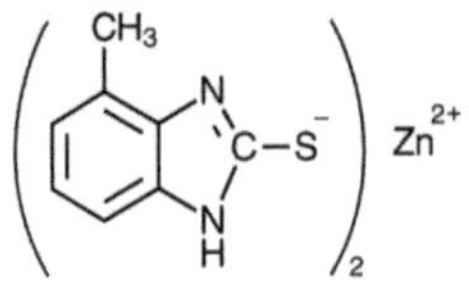

Figure 1.38: ZMTI.

1.7 Antiozonants

It is sometimes necessary to compound latices to resist ozone, especially products made from unsaturated polymers such as NR. The usual route is to mix a paraffin wax into the formulation, especially if an amine antioxidant cannot be used. The wax will be added into the compound as an emulsion. As with other emulsions, the wax can be bought in ready emulsified, or prepared in house, usually by melting the paraffin wax and using one of the methods for emulsification described in Chapter 2. After dipping, the wax will subsequently bloom to the surface of the polymer and provide an ozone-resistant layer. The rate of blooming is dependent on the temperature and the MW distribution and solubility of the wax in the polymer. As the wax on the surface is removed through wear, further wax can bloom to the surface, as long as sufficient wax remains in the bulk of the rubber to be above its solubility limit. Specially formulated blends of wax are commercially available to offer protection over a range of storage and service temperatures. The level of addition of wax to rubber will depend upon the type of wax, the type of polymer and the application, but will usually be in the range 1 to 4 or 5 phr, although higher levels may be required in specific circumstances. It will be important to balance the concentration of wax required to give adequate ozone protection against any loss of physical properties that may result from wax addition. Another route to confer increased ozone resistance is to laminate a skin of an ozone-resistant polymer, such as CR, to the article, although there will obviously need to be sufficient compatibility between the skin and the substrate to ensure adequate interlaminar adhesion.

1.8 Fillers and pigments

1.8.1 Fillers

Fillers are added to latices to reduce costs or enhance properties through reinforcement. There are many different materials that have been used in latex systems over the years, with varying degrees of success. A more detailed treatment of fillers, their use in dipped latex formulations and their reinforcing effect (if any) is given in Chapter 5. The following is a list of some of the materials that have been more commonly encountered or proposed as fillers for latex systems:

- Calcium carbonate, also as whiting or chalk.
- Clays of many different types.
- Blanc Fixe [barium sulfate ($BaSO_4$)].
- Lithopone [a mixture of $BaSO_4$ and zinc sulfide (ZnS)].
- Silica, fumed, precipitated, and colloidal and other silicates.
- Formaldehyde condensation resins.
- Polymer latices such as high-styrene SBR.
- Dispersions of polymers, such as polystyrene, polyvinyl chloride, PMMA, PU, polyvinyl acetate, polyacenaphthylene and so on.
- Graft polymers.
- Modified starch.
- Proteins.
- Natural polymers such as lignin, chitin, chitosan and shellac.
- Carbon nanotubes and other carbon structures such as buckminsterfullerene.

Adding finely divided materials such as fillers into latices can pose stability problems, as the new surface of the filler particles competes with the latex particles for surfactant and other stabilising species. When adding fillers it will be necessary to take this into account, and ensure that adequate stabiliser is present to maintain the stability of the system.

1.8.2 Pigments and dyes

Many dipped latex articles will be coloured and there are certain recommendations that should be followed in this respect. Pigments are generally preferred, as they tend to be more colour-fast and more resistant to extraction, although both can be encountered in latex formulations. Pigments are added as a dispersion of the solid colourant, whilst dyes are generally added as a solution. First of all the colourant must be compatible with the latex. This might sound to be obvious, but the author has known of pigment dispersions with a pH of around 2 being used in NR latices, which, of course, will lead to problems with coagulum. Many dipped latex products are used as medical devices or other controlled applications such as food handling. In these cases both the pigment or dye, and any other component used in the preparation, must be suitable for the application. Advice should be available from the supplier, and the regulatory status of the components may be available from other published sources, such as the US Food and Drugs Administrations' Codes of Federal Regulations.

As well as factors of the dispersion or solution, such as the pH, the chemical nature of the colourant should be taken into account. Many types of latices are sensitive to heavy metal ions and so any pigments based on, for example, iron, chromium or zinc salts must be thoroughly evaluated for compatibility before use. Bear in mind that the release of metal ions from some of these compounds may be slow, and any stability problems associated with their use may take some time to become evident.

Luminous pigments – generally phosphorescent pigments – in particular can cause problems. The older material used to give phosphorescence was copper-doped ZnS. Both copper and zinc can cause problems when used in latices: copper ions can promote the oxidative degradation of polymers, such as NR latex, and zinc ions can promote latex instability. In addition, ZnS has a density of around 4 g/cm^3 and can therefore be difficult to keep in suspension. The more modern phosphorescent pigments, which give a much more intense and longer lasting luminescence, are based on strontium aluminate, doped with one of the rare earth metals such as europium or dysprosium. These materials, as well as having a similar density to ZnS, can slowly release heavy metal ions and NR latices compounded with them will gradually thicken over a period of a few days.

As mentioned above, pigments tend to be more stable and resistant to extraction than dyes. However, pigmented dipped rubber articles can bleed colour under certain conditions, for example, with a high pigment loading or in an alkaline environment. It has also been noted that high drying temperatures can increase the tendency for coloured NR articles to bleed colour.

1.9 Testing of raw materials

The testing of raw materials should, as far as possible, be targeted and sensible, with focus on those materials deemed as critical to quality and important for the end use. Once the appropriate tests have been selected, it is important to ensure that the tests are carried out correctly, using well-trained staff, robust and validated procedures, and correctly calibrated equipment. Above all, make sure that the results generated by the tests are used. Testing costs money and should be viewed as insurance against material-related problems: merely filing the test results away is not the best use of that investment. It is recommended that the results of goods inward tests are reviewed regularly – graphing them is a good idea – so that any changes or trends can be quickly picked up.

1.9.1 Latex

There are many tests and test methods for latices listed in the ISO standards catalogue, although far fewer will be encountered in the day-to-day operation of a dipped latex goods factory. The most commonly encountered ones are likely to be those listed in Table 1.4 (the versions are correct at the time of writing).

Table 1.4: Latex test methods.

Property	ISO standard
Mechanical stability – NR latex	ISO 35:2004
Total solids content	ISO 124:2014
Alkalinity – NR latex	ISO 125:2011
Dry rubber content – NR latex	ISO 126:2005
KOH number – NR latex	ISO 127:2012
Volatile fatty acid number – NR latex	ISO 506:1992
Coagulum content	ISO 706:2004
Determination of pH	ISO 976:2013
Viscosity – Brookfield method	ISO 1652:2011
Specifications for NR latex concentrate (ammonia preserved, centrifuged or creamed)	ISO 2004:2010
MST – synthetic rubber latex	ISO 2006:2009 (Parts 1 and 2)
CR latex – determination of alkalinity	ISO 13773:1997

Full descriptions of how to carry out the tests are described in the relevant standards and will not be covered in detail here. However, it is worth bearing in mind the following possible common errors when applying or interpreting these tests.

1.9.1.1 Mechanical stability time

If the test results are likely to be compared against those from the latex supplier then the test conditions in the international standard need to be followed exactly – that is what the supplier will be doing. For example, for NR latex the test should be carried out at a temperature of 35 ± 1 °C, with the latex adjusted to a total solids content of 55 ± 0.2%. It can sometimes be difficult to determine the exact end-point in this test, so it is always worth continuing the test for another 15 seconds or so to confirm that the end-point has been reached. Most models of test machines for measuring MST already have verification of the correct speed – 14,000 rpm – built in, through vibrating reeds or lights, but if it is required to verify the speed, this must be done with a non-contacting tachometer. Contacting tachometers will slow down the rotation and give false readings.

1.9.1.2 Measurement of pH

pH meters should be calibrated regularly – ideally daily – using the procedures in the manufacturer's instructions. The calibration should use at least two points, one using a buffer solution of a nominal pH of 7 and the other using a buffer solution corresponding to the approximate pH of the sample to be measured. For most latices used in dipping, this second pH will normally be around 9 or 10. Buffer solutions are commercially available or can be made up using one of the several formulations available in the literature. Note that the samples of buffer solutions used to calibrate the pH meter should be discarded after use and not reused. Note also that alkaline buffer solutions are not stable and will eventually absorb enough atmospheric carbon dioxide to alter their value. It is necessary to maintain them in a firmly closed container and observe the shelf life. It is worth recording the details of the regular calibration, including, where available, any value of the electrode gradient (slope). Any drift or changes in the slope outside of the manufacturer's recommended limits can indicate the need for maintenance or replacement of the electrode. Always store the electrode in contact with the solution recommended by the manufacturer: never store it dry or in contact with distilled or demineralised water.

1.9.1.3 Measurement of viscosity

Viscosity can be measured by several methods, but the most common ones are using a Brookfield viscometer or a flow cup. The Brookfield method is recommended for raw materials, as it is likely that this will be the method used by the supplier. In both cases the equipment needs to be matched to the viscosity by using the correct choice of spindle and speed of rotation for the Brookfield viscometer, and correct choice of orifice for the flow cup. Viscosity is highly dependent on temperature, and the temperature of the latex must be controlled and reported together with the viscosity results. The test temperatures in the ISO 1652:2011 [22] standard are 25 ± 2 °C or 27 ± 2 °C in tropical countries. If a flow cup is used, it is very important that the latex is filtered before testing and that there are no air bubbles in the sample.

1.9.2 Other raw materials

Other raw materials – accelerators, surfactants, antioxidants and so on – will usually be bought to a certificate of analysis (COA). In the case of a reputable supplier, the COA may be sufficient guarantee of quality, however, it is common practice to carry out a few confirmatory tests for identity and purity. A visual examination should check that the packaging is undamaged and that the material is free from dirt and other contamination. For materials such as accelerators or antioxidants a simple melting point can serve as a guide to identity and purity, and other tests that may prove useful are infrared spectroscopy and ultraviolet spectrophotometry, where the results can be compared against those from a reference standard.

1.10 Storage of raw materials

The raw materials for latex compounding should be handled and stored carefully, as some of them can be dangerous if incorrectly handled. In particular, the relevant MSDS should be read by all personnel who may be involved with the materials and copies of the MSDS (in the local language if necessary) should be displayed.

1.10.1 Latex

All latices are two-phase systems in a state of metastable equilibrium. Energetically, they would rather exist as a block of rubber and a puddle of water, which means that latices must be stored and handled correctly. Extremes of temperature must be avoided – ideal storage temperatures are between 10 and 25 °C. In tropical climates latex should be stored under cover, away from direct sunlight, and in colder climates storage should be indoors, to avoid the risk of freezing. In general, latices should be agitated periodically, to ensure that they remain homogeneous. The specific gravities of the elastomers commonly encountered in dipping are shown in Table 1.5. As can be seen, the rubber phase in both NR and synthetic polyisoprene latex is lighter than water, and if left unagitated the rubber particles will rise to the top – known as creaming. The cream can usually be redispersed by agitation, but it is better to prevent the creaming by regular agitation. This is less important with NBR or CR latices, as the smaller particle size helps maintain homogeneity. Nevertheless, it is recommended that all latices are agitated before use, which can be accomplished by mechanical stirrers, or if the latex is in drums, by rolling the drums. Note that any agitation must be gentle – excessive agitation can cause coagulation.

Table 1.5: Specific gravities of some latices used for dipping.

Elastomer	Specific gravity
NR	0.92
IR	0.92
NBR	Approximately 1.0, but varies with composition
CR	1.20–1.24

Pumping latices can also cause coagulation unless the correct design of pump is used. Diaphragm pumps or peristaltic pumps are usually preferred, as they have a history of successful use with latices. It is common practice to transfer latex initially to an upper story, either for storage or compounding, so that all subsequent transfers can be made by gravity, reducing the need for pumping.

1.10.2 Other raw materials

In general, all raw materials should be stored in a dry, well-ventilated area, avoiding extremes of temperature. There are some specific requirements for certain raw materials, as detailed below.

1.10.2.1 Sulfur

Sulfur is classified as a dangerous material for transportation, but the usual form in which it will be encountered in bulk in a latex dipping factory will be as the loose powder or in bags. Sulfur is flammable and the single biggest risk is that of a dust explosion. Dust can be produced when the bags are handled at storage and when the sulfur is weighed out at the compounding stage. Of these, by far the bigger risk relates to storage procedures. Sulfur dust tends to persist in the air and in a confined space the concentration may approach the lower explosive limit of approximately 35 g/m^3. As it very difficult to avoid dust when bags of sulfur are handled, it is recommended that all potential sources of ignition are excluded in areas where sulfur is stored. This will include taking appropriate measures with electrical equipment, enforcing no smoking regulations and ensuring that electrical earthing is present at any point where static electricity could build up. Any damage to bags should be repaired if possible, or the bag placed in an intact outer bag. The aim should be to minimise any dust in the environment and it will help to reduce the risk if the sulfur is stored in a separate area. Bags of sulfur, in common with other powder materials, can be contaminated on the outside with dust, which will make it very difficult to attach adhesive identification labels showing the status of the material. In these cases it may be necessary to staple the labels to the ends of the bags or wipe off external dust before applying the labels.

1.10.2.2 Accelerators

Accelerator powders, such as ZDEC or ZDBC, can also pose risks of a dust explosion and, as with sulfur handling, procedures should minimise dust formation and potential sources of ignition should be excluded. In addition, most rubber accelerators have a degree of toxicity and personal protective equipment such as dust masks, eye protection and appropriate protective gloves should be worn.

1.10.2.3 Ammonia

Ammonia – specifically NH_4OH – is supplied for use as a solution in water. In Europe it is often encountered as a 35% m/m solution (often known as 0.880 or '88' ammonia after its density) but in warmer countries it will be supplied as a more dilute solution, for example, 28 or 25% m/m. NH_4OH is a corrosive alkaline liquid with an intensely irritating odour. It can burn skin and is very dangerous if it comes into contact with the eyes, so adequate personal protective equipment must be worn when handling it. It is recommended that NH_4OH is stored within a bund wall (a low wall to contain any unintended spillage or escape of the material) and in a well-ventilated area, so that any leaks or spills have a limited effect on the rest of the facility.

1.10.2.4 Acids

Acids are frequently used in latex dipping factories to clean formers and to coagulate waste latex. As with ammonia, it is recommended that they are stored behind a bund wall, well away from any alkalis or other materials with which they could react violently. As with NH_4OH, adequate personal protective equipment must be worn when handling acids.

1.10.2.5 Other raw materials

As mentioned above, other raw materials should be stored in a dry place, avoiding extremes of temperature. All raw materials should be well and clearly labelled as to status, taking care that labels remain affixed to bags and containers.

1.10.2.6 Time-expired materials

All raw materials will have a shelf life, and the production and reorder schedules should be geared to use up the materials before the expiry of the shelf life. However, there will be times when this might not be possible, and it is recommended that the factory has a procedure in place to deal with time-expired materials. These procedures will depend on the materials in question, but a suggested approach would be to test the materials to the original purchasing inspection, and if they still comply with the requirements assign another shelf life of 50% of the original shelf life. Of course any such procedures would need to be fully validated to confirm that the quality of the product is unaffected.

References

1. D.C. Blackley in *Polymer Latices: Science and Technology*, Volume 1, Chapman & Hall, London, UK, 1997.
2. *Natural Rubber Science and Technology*, Ed., A.D. Roberts, Oxford University Press, Oxford, UK, 1988.
3. D.C. Blackley in *Polymer Latices: Science and Technology*, Volume 2, Chapman & Hall, London, UK, 1997.
4. ISO 2004:2010: Natural rubber latex concentrate – Centrifuged or creamed, ammonia-preserved types – Specifications.
5. A. C-P. Loh in *Further Investigations of the Prevulcanization of Natural Rubber Latex*, The North London Polytechnic, London, UK, 1982 [PhD Thesis].
6. A. Kells and R. Groves in *Proceedings of Latex 2006*, Frankfurt, Germany, 2006, Paper No.16.

7. *The Vanderbilt Latex Handbook*, Ed., R.F. Mausser, R.T. Vanderbilt Company, Inc., Norwalk, CT, USA, 1987.
8. C. Petri in *Proceedings of Latex 2004*, Hamburg, Germany, 2004, Paper No.3.
9. J. Tinkler, D. Gott and J. Bootman, *Food and Chemical Toxicology*, 1998, **36**, 849.
10. M.W. Philpott in *Proceedings of the 4th Rubber Technology Conference*, London, UK, 1962, p.470.
11. C. Webster and C. Nola in *Proceedings of Latex 2006*, Frankfurt, Germany, 2006, Paper No.9.
12. A.D.T. Gorton, *NR Technology*, 1988, **19**, 61.
13. C. Hepburn, *Rubber Technology International*, 1985, **24**, 5.
14. J.C. Carl in *Neoprene Latex: Principles of Compounding and Processing*, E.I. Du Pont de Nemours & Co. Inc., Wilmont, DE, USA, 1962.
15. M.E. Cain, G.T. Knight, P.M. Lewis and B. Saville, *Journal of the Rubber Research Institute of Malaya*, 1969, **22**, 289.
16. M.E. Cain, K.F. Gazeley, I.R. Gelling and P.M. Lewis, *Rubber Chemistry and Technology*, 1972, **45**, 204.
17. M.E. Cain and B. Saville, inventors; The Natural Rubber Producers' Research Association, assignee; BP1185896, 1970.
18. G. Scott in *Developments in Polymer Stabilization – 4*, Ed., G. Scott, Applied Science Publishers Ltd., London, UK, 1981, Chapter 6.
19. J.A. Kuczkowski and J.G. Gillick, *Rubber Chemistry and Technology*, 1984, **57**, 621.
20. *Developments in Rubber* Technology *1: Improving Product Performance*, Eds., A. Whelan and K.S. Lee, Applied Science Publishers, London, UK, 1979.
21. C. Webster and C. Nola in *Proceedings of Latex 2004*, Hamburg, Germany, 2004, Paper No.4.
22. ISO 1652:2011: Rubber latex – Determination of apparent viscosity by the Brookfield test method.

2 Latex compounding

2.1 Introduction

The process of adding the various components of the formulation to the latex and preparing the latex for dipping is called compounding. It may or may not include prevulcanisation, depending on the latex and the manufacturing process to be employed. Compounding is a critical stage – if not the most critical – in the manufacture of dipped latex articles, and if not performed correctly will have a major impact on quality. On the other hand, good and consistent compounding procedures can make subsequent manufacturing operations much easier. One may make bad products from a good latex compound, but it is very hard to make good products from a bad latex compound. The practice of compounding is simple: understanding the principles behind it less so. One of the key factors to remember is careful attention to detail. A compounding procedure can be fundamentally excellent, but if it is not rigorously followed then the characteristics of the resulting latex will vary, which in turn will translate into variability in the dipped products. Similarly, training of the operators involved in compounding is important, not just in the processes employed in the procedure, but also the reasons behind the procedures.

A good-quality compounded latex will have the following characteristics:

- Consistent properties.
- Stability appropriate to the end usage.
- Conformance to the compounded latex specification.
- Satisfying the customer's needs.

It will also have been produced in an environmentally sound and energy-efficient manner.

The ideal latex formulation will also be simple, containing no more ingredients than are necessary. The author has seen many cases where the formulation has 'grown' over the years, as additional components have been added, either to solve problems or because someone thinks it is a 'good idea' – possibly following the visit of a sales representative. Much advice on compounding will be given in this chapter, but if your procedures work for you then don't change them just for the sake of it. However, one more piece of advice: don't be tempted to try and save money through using cheap raw materials, especially latex. Compounding latices can give enough problems without adding more. Buy from reputable suppliers – they will generally have experienced and knowledgeable staff to provide technical back-up. Be aware of expiry dates and have written and validated procedures to deal with time-expired materials.

The compounding stages can be summarised as follows:

- Raw material preparation.
- Addition of ingredients.

https://doi.org/10.1515/9783110638097-002

- Prevulcanisation [predominantly with natural rubber (NR) latex].
- Maturation.
- Final compounding.
- Testing and monitoring.
- Clean-up.

2.2 Compounding equipment

Suitable mixing and storage vessels should be used for latex compounding. Virtually all types of latex for dipping are alkaline [especially polychloroprene (CR) latex] and the compounding vessel, and all surfaces exposed to the latex, should be capable of resisting alkalis. The containers, pipes and pumps should be easy to clean, preferably using the minimum amount of water, as many dipping plants are located in areas where water may be scarce or expensive.

If it is intended to prevulcanise the latex at an elevated temperature then some form of heating of the latex in the compounding tank is required. Such heating should be gentle, as latices are temperature-sensitive materials. A jacket around the mixing vessel, through which hot water (say at 60 °C) can be circulated, is ideal. Heating media such as live steam should be avoided, as the large temperature gradients that will result can cause coagulation of the latex. With appropriate plumbing the same jacket can be used to circulate cold water around the vessel to cool the latex when required.

The latex will need to be agitated during compounding and agitation must ensure that the compounding ingredients are dispersed throughout the latex, sedimentation of the denser materials is prevented and equilibration of temperature is maintained. However, latices are shear-sensitive materials and excessive agitation must be avoided, as must the incorporation of air. With these things in mind, the ideal agitation will use large stirrer blades moving relatively slowly – for example, between 40 and 60 rpm – through the latex. Compounding also requires calibrated balances of suitable capacity, and several calibrated jugs and buckets for ease of handling smaller quantities of materials.

It is recommended that the latex and compounding ingredients are filtered either before or during addition to the latex. This operation is easier if there is some kind of filter assembly that can be easily incorporated into the material flow. Either disposable filter media – such as a filter cloth – or a reusable medium – such as a stainless steel mesh or Nylon filter cloth that can be cleaned after use – is suitable. If the latex is to be filtered through an absorbent material such as cheesecloth, it is recommended to soak the cloth in ammoniated water first, so that the cloth does not absorb water from the latex on first contact, which may cause the latex to coagulate and block the filter. The mesh size needs to be a compromise: too large a mesh size will be ineffective and too small a size will be slow and block easily; typically a mesh size of around 80 (approximately 180 μm) is used. To weigh out the materials, balances of appropriate range and accuracy will be needed. Balances with printing capability will provide a printed record of the quantities, which can be attached to the compounding record sheet. Finally, the quantity of latex needs to be measured. An inline flow meter in the latex feed pipe can be used, or the quantity verified in the compounding vessel with a calibrated dip-stick. Strain gauges positioned under the compounding vessel can also be used to weigh the latex. Whatever method is used to measure the quantity of latex, it is recommended that the method is validated.

2.3 Raw material preparation – liquid materials

The compounding ingredients must be added in a form that is compatible with the latex. In general, this means as a water-based liquid with a pH and ionic strength that will not cause instability in the latex.

Liquid components should not usually be added without dilution and adjustment of ionic strength and pH. Water-miscible materials will be mixed with water, whilst immiscible liquids will usually be emulsified. It is often possible to buy the raw materials ready emulsified, and this can be a very convenient way to obtain high- quality stable emulsions, although, as with dispersions, the full formulation may not be disclosed. There are several ways of preparing emulsions inhouse, of which the following are the most often encountered:

- The liquid to be emulsified can be carefully stirred into a solution of surfactant in water, ideally using a high-shear mixer. The surfactant solution may give rise to a high level of foam during the initial stages of preparation, so it may be better to start off with a more dilute surfactant solution and add a more concentrated solution as the mixing proceeds.
- The liquid to be emulsified can be mixed with the surfactant and then mixed into the water, again using a high-shear mixer.
- In some cases it can be convenient to form the surfactant *in situ* as the liquid is mixed into the water. For example, if the emulsifying surfactant to be used is a fatty acid soap, the fatty acid can be dissolved in the liquid to be emulsified and potassium hydroxide (KOH) dissolved in the water. Then, when the liquid is mixed with the alkaline water, the fatty acid will react with the KOH to form the soap and stabilise the emulsion.
- Finally, in some cases it may be possible to disperse the surfactant solution into the liquid to form a water-in-oil emulsion which will then phase invert as more water is added.

High-shear mixing will be required in all cases along with, in some instances, experiments in the laboratory to identify the best method for a particular ingredient. As noted above, attention should be paid to concentration and pH.

Several different types of surfactant solution can be used to stabilise emulsions, including non-ionic surfactants (for example, ethylene oxide condensates) or fatty acid soaps. In general, those surfactants that have a lower tendency to foam are preferred.

2.4 Raw material preparation – solid materials

Solid materials will be added as a solution or, in the case of materials with no or little water solubility, as a dispersion. They must not be added as dry solids – this will almost certainly lead to the formation of coagulum. Solutions of compounding ingredients are simple to prepare using standard mixing technology, whereas dispersions can be more difficult.

High-quality dispersions of commonly used compounding ingredients are readily available from specialist companies. Buying in these materials is easier, but generally more expensive than preparing the dispersions inhouse. In addition, all of the details of the dispersion formulation may not be disclosed by the supplier, and the dipping company may want to keep their latex formulation confidential. For these reasons many companies prepare their own dispersions.

Dispersions are usually ground (or 'milled') to a small particle size – say less than 5 µm although this small size is not necessarily for reactivity with the latex. Experiments involving the prevulcanisation of NR latex have shown that the rate at which the rubber particles react with sulfur and accelerators is independent of the particle size of the dispersion [1]. In fact, prevulcanisation of NR latex has been shown to occur when the sulfur or accelerator was presented to the latex as a coating on the inside of the prevulcanisation vessel [2]. The reasons for using a small particle size are to reduce sedimentation of the active ingredient, helping to keep it in dispersion, and to reduce the likelihood of any large particles of material remaining in the finished rubber film which could initiate failure. A smaller dispersion will also improve clarity of the finished product. It should also be noted that some data suggest that the protection offered by a phenolic antioxidant is improved by the use of a superfine dispersion [3]. In some situations however – for example, where a compounded latex is subsequently centrifuged or allowed to sediment to remove solid residues – it may be advantageous to use a dispersion of a larger particle size, which is more easily separated. Note that residual dithiocarbamate accelerators are powerful antioxidants, and removing them may adversely affect ageing performance. If removal of these residues is desired then the effect on ageing should be assessed and antioxidant protection enhanced if necessary.

Dispersions are usually prepared by grinding the solid material mixed with water. This grinding can have two consequences: firstly, aggregates of fine particles will be broken up, but there is no effect on the primary particle size. This deagglomeration is a relatively low energy operation, and can be achieved by, for instance, high-shear mixing. However, in many cases it will be necessary to reduce the size of the primary particle of the material in order to obtain a dispersion of the required particle size. This operation requires much more energy and will take place in, for example, an attrition mill ('attritor'), a ball mill, a bead mill or one of the new high-energy mills.

The duration of milling is important. If the milling operation is continued for too long, the material can start to reagglomerate and the particle size will increase. Examples of what can happen if the milling time is continued too long are shown in Figures 2.1 and 2.2.

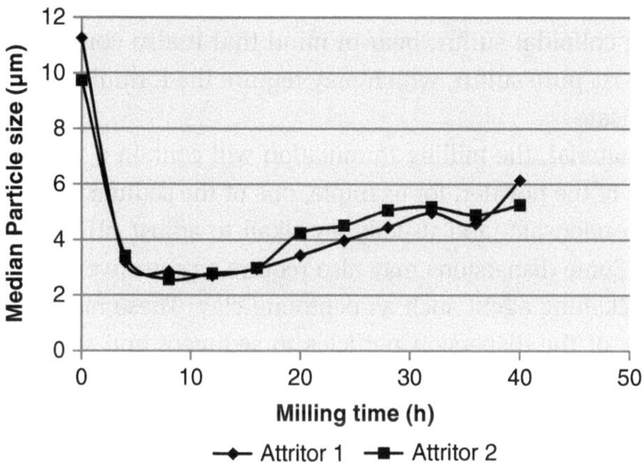

Figure 2.1: Changes in particle size of an accelerator dispersion with time in an attrition mill.

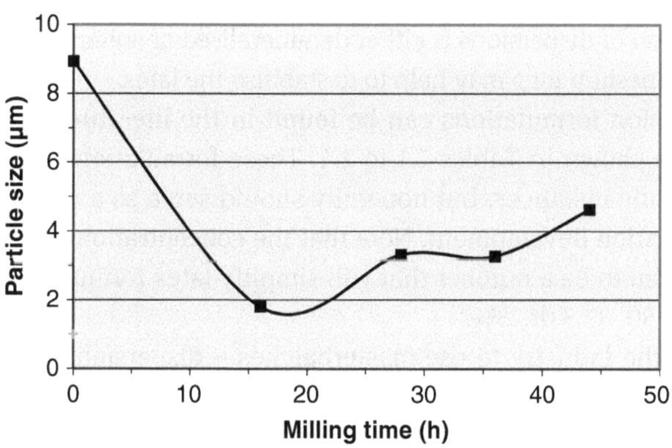

Figure 2.2: Changes in particle size of a sulfur dispersion with time in an attrition mill.

In Figure 2.1 an accelerator dispersion was milled in an attritor for up to 42 h. It can be seen that the median particle size (measured using a light-scattering method) reached a minimum after about 8 h, and then slowly increased thereafter, most likely due to the reagglomeration of particles. A repeat experiment in another attritor gave almost identical results. Any milling after 8 h in this example is a waste of time and energy. Figure 2.2 shows the change in particle size of a sulfur dispersion, again milled in an attritor and measured using a light-scattering method. The same pattern can be seen. In this case the optimum milling time appears to be between 15 and 20 h. It is important in any milling operation that the conditions – formulation, viscosity, time of milling and so on – are optimised, not just to obtain the best possible dispersion but also to ensure that time and energy are not being wasted.

It is worth mentioning here that sulfur is available – at extra cost – in a colloidal form, which requires much less in the way of milling to give a dispersion of acceptable particle size. If using colloidal sulfur, bear in mind that it also contains a dispersing agent and is not 100% pure sulfur, which may require the formulation to be adjusted slightly to compensate.

In addition to the active material, the milling formulation will contain a wetting agent to assist in rapid wetting of the powder, for example, one of the sodium salts of a naphthalene–sulfonic acid condensate, and usually an alkali to adjust pH and to help prevent reagglomeration. Some dispersions may also require a protective colloid such as a caseinate, and a thickening agent such as bentonite clay. These materials can help counter the tendency of the dispersion particles to sediment and stick together. However, as with latex formulations, it is best to keep the dispersion formulation as simple as possible, with ingredients only added when they are known to be necessary. Surfactants may be used in the grinding stage to help prevent reagglomeration, but are also often added later, as milling in the presence of a surfactant can generate an unacceptable level of foam – and as will be discussed later, foam can inhibit grinding. As with the latex formulation, advice regarding the optimum surfactants for these purposes can be obtained from the surfactant suppliers. It is recommended that the water used in the preparation of dispersions is either demineralised or softened, as the presence of calcium or magnesium ions may help to destabilise the latex.

Many examples of dispersion formulations can be found in the literature and some typical formulations are shown in Tables 2.1 to 2.4. These formulations may need to be fine-tuned for specific instances, but hopefully should serve as a convenient starting point for formulation development. Note that the concentration of active ingredient is usually chosen to be a number that will simplify latex formulation calculations, for example, 60, 50, 33.3 or 25%.

It is common practice in the industry to use masterbatches – dispersions containing more than one active ingredient – which undoubtedly simplifies compounding. However, a milling formulation works best if it is optimised for the equipment and materials. As mentioned earlier, zinc oxide (ZnO) is an easy material to mill,

Table 2.1: Examples of formulations for sulfur dispersions.

Chemical	50% Sulfur dispersion	60% Sulfur dispersion
Sulfur	50 wt%	60 wt%
Wetting agent*	0.8	1.0
Bentonite clay	0.4	0.5
Casein	0.2	0.2
Biocide	0.01	–
25% Ammonia solution	–	0.05
Water	To 100%	To 100%

*e.g., the sodium salt of a naphthalene–sulfonic acid condensate

Table 2.2: Examples of formulations for accelerator dispersions.

Chemical	Dispersion 1	Dispersion 2
Accelerator	50 wt%	50 wt%
Wetting agent*	1.0	2.0
Bentonite clay	0.4	0.5
Casein	0.1	–
25% Ammonia solution	–	0.05
Water	To 100%	To 100%

*e.g., the sodium salt of a naphthalene–sulfonic acid condensate

Table 2.3: Examples of formulations for ZnO dispersions.

Chemical	50% ZnO dispersion	55% ZnO dispersion
ZnO	50 wt%	55 wt%
Wetting agent*	1.0	2.0
Bentonite clay	0.5	–
25% Ammonia solution	–	0.25
Water	To 100%	To 100%

*e.g., the sodium salt of a naphthalene–sulfonic acid condensate

whereas sulfur or some antioxidants are much more difficult to reduce to a small particle size. If your masterbatch contains both ZnO and sulfur, it will be difficult, if not impossible to set up milling conditions that will deal with both materials effectively at the same time. If the masterbatch approach is required it is much better to mill separately and then combine.

Table 2.4: Examples of formulations for antioxidant dispersions.

Chemical	50% Antioxidant dispersion	33.3% Antioxidant dispersion
Antioxidant	50 wt%	33.3 wt%
Wetting agent*	2.0	1.0
Bentonite clay	0.4	0.25
Nonionic surfactant**	0.25	–
25% Ammonia solution	0.05	0.05
Water	To 100%	To 100%

*e.g., the sodium salt of a naphthalene–sulfonic acid condensate
**e.g., a fatty alcohol ethoxylate

The grinding operation should usually be carried out using as high a concentration as is practical, as grinding is more efficient at high viscosity where the shear stress is higher. The choice of equipment will usually depend upon what is available, but all types of comminution equipment currently used in the latex industry can generate dispersions of an adequately small particle size. The main types of milling equipment used in the industry are the ball mill, the attrition mill ('attritor'), the vibromill, and the bead or sand mill. There are also planetary mills that are very efficient at giving a fine dispersion quickly, but these are in general too small a capacity for production use. It is important that none of these types of milling equipment are run dry, as this will promote excessive wear of the milling media and also the interior of the machine.

2.4.1 Ball mill

The traditional ball mill (or pebble mill) is a rotating cylinder containing steel or ceramic balls, pebbles or cylinders. When it is rotated, the balls or other milling media cascade down, and any other materials present, such as a dispersion, are subject to impact and shear forces by the milling media. To assist in the lifting and cascading of the milling media, baffles can be fixed to the inside surface. A ball mill has the advantage that no premixing of the ingredients is necessary, as this will take place as soon as the mill is started. The system is closed, so there is no material loss and the technology is simple, requiring no special skills and minimum maintenance. A ball mill is efficient, can work at a high solids concentration and is available in a wide range of sizes. They can run with minimum supervision, be standardised and scale-up from laboratory-sized mills to larger production machinery is usually straightforward. However, ball mills are noisy and slow compared with some of the other technology available today. It is important to ensure that the loading of grinding media and dispersion into the ball mill is correct so that milling efficiency is maximised. The speed of rotation of the mill should be fast enough to energise the milling media, so that the balls are lifted up and can then cascade down through the dispersion. If the rotation is too fast the balls will be thrown around inside the drum ('cataracting'), which increases the wear of the balls and the drum lining, rather than reducing the particle size of the dispersion. If the speed of rotation is even faster, the balls will be held against the inside of the drum by centrifugal force and will not result in any milling at all. The speed at which this happens is called the 'critical speed' and can be calculated using the formula:

$$\text{Critical speed (rpm)} = 42.29/D^{1/2} \tag{2.1}$$

where 'D' is the internal diameter of the drum in metres.

The optimum speed for cascading is generally considered to be 50 to 75% of this critical speed, although in many cases the best conditions will be determined by practical experience.

For the most efficient performance, the milling media should occupy half the volume of the ball mill, as this will allow the balls to cascade freely across the full diameter of the mill. The charge of dispersion should be sufficient to just cover the balls. If more dispersion is present, it will take longer to achieve the required particle size reduction; if less is present, then wear of the balls will be increased. However, there may be occasions when the time taken to achieve the required particle size is less important than the batch size, and in these cases it could be better to increase the volume of dispersion and accept the longer milling time. As with so much in dipped product manufacturing, decisions should take practical considerations into account, as well as theoretical ones.

There is, and probably always has been, debate regarding the optimum type and size of the milling media to be used. Ball density is obviously important, as is the composition. Porcelain media, which have a density of around 2.5 g/cm^3, are widely used, and are available in a range of shapes and sizes. Other ceramics, such as those based on zirconium oxide, are also popular and can have densities up to around 6 g cm^3. In practice it is found that a smaller ball gives a finer dispersion. Although a larger ball has more mass, the smaller balls will give rise to more impacts in unit time, and the consequence is that the grinding will be more efficient and faster. However, using too small a ball size will give rise to problems emptying the mill and separating the balls from the dispersion. It is common practice to use a mixture of ball sizes, to take advantage of the greater mass of a larger ball and the increased frequency of impact of the smaller balls, but this is considered by some to be ineffective and the larger balls merely wear out the smaller balls faster [4] and use more energy in the process. The author has not found any advantage of using a mixture of ball sizes. The optimum shape of the grinding media is another contentious area, with spheres, cylinders and irregular shapes all being available. It has been shown [4] that the more varied impacts available by using cylinders (i.e., point, line, edge and face contacts) give a narrower particle size distribution than the point contact alone achieved using only spherical media.

Whatever type of milling media is used, don't forget to regularly check for wear, and replace when necessary. Replacement with fresh milling media has been shown to give an immediate and substantial improvement in the quality of dispersions, which in turn can lead to improvements in dipped latex products. The media may not be cheap, but if the product yield increases only fractionally, the cost may soon be recovered.

Milling media are subject to wear during a milling operation, and any material worn from the media will become part of the dispersion. In a properly run system the amount of wear will be very small and contamination from ceramic media can usually be ignored. However, owing to the negative effects that iron can have on rubber latices, it is recommended that steel balls are not used to prepare dispersions for use in the latex industry.

As mentioned above, the viscosity of the dispersion is important in order to obtain the most effective particle size reduction, which is achieved *via* a combination of shear and impact. A high viscosity will result in high shear stress during milling, but a high viscosity can also cushion the impact of the balls and reduce the effect of the impact. If the viscosity is too low there will be little contribution from shear in the milling process. Note also that the viscosity will fall if the dispersion heats up during milling. If the dispersion foams during the milling, the foam can cushion the impact of the milling media and reduce milling efficiency. Each type of equipment will have an optimum viscosity that needs to be established and then used. Further details on the use of ball mills are readily available in the literature.

2.4.2 Attrition mill

The attrition mill, or attritor, is an upright container, filled with milling media such as pebbles or balls. A vertical shaft, fitted with hardened steel fingers that extend almost to the walls of the vessel, is mounted in the centre of the vessel and connected to a powerful motor. In use, the rotation of the centre shaft powers the fingers through the milling media, transferring substantial amounts of energy into the media and the dispersion. Sufficient dispersion is usually added just to cover the milling media, which will generally be in the form of pebbles or ceramic balls. Like the ball mill, it is not necessary to premix the dispersion components, and it can work at a high solids concentration. An attritor can reduce the particle size much quicker than a ball mill and can similarly work at a high solids concentration. However, the sheer amount of energy that an attritor puts into a dispersion means that heat build-up can be a problem, and cooling is essential. They are also noisy machines in use and are ideally situated in an enclosed environment. Another potential problem of the attrition mill is that the motor and gearbox are generally positioned above the dispersion, and any oil leakage gives a potential risk of contamination. For this reason a regular preventative maintenance programme is essential. Attritors are generally fitted with a pump to allow the dispersion to be circulated if required to eliminate the effect of any dead spots.

2.4.3 Vibration mill

The vibration mill ('Vibromill') is another similar type of mill commonly used to reduce the particle size, and many of the comments made about the milling media in the section on ball mills will apply. It consists of a chamber containing the dispersion and the milling media, mounted on springs and connected to a motor. The motor allows the container to be agitated at high frequency. The milling media generally used are either pebbles, ceramic balls or short ceramic cylinders. Like the ball mill, it is a closed system that can run with minimum attention, although it is generally better to premix the dispersion before it is milled. A Vibromill is an efficient method of reducing the particle size of dispersions, although it is not as fast as the attrition mill.

Whilst the ball mill is best run half-full, to maximise the grinding efficiency, the attrition mill and vibration mill can be run with the chamber full, resulting in better utilisation of the chamber volume. Like the ball mill, and the attrition mill, the vibration mill is a batch process.

2.4.4 Bead mill/sand mill

Bead or sand mills are becoming increasingly used nowadays, and are capable of producing small particle size dispersions quickly and efficiently. They comprise a cylinder (which can be vertical or horizontal) through the centre of which runs a shaft containing a number of discs, usually studded with holes or slots. The discs are a close fit to the cylinder, and can be rotated at high speed – typically several thousand rpm. The chamber is virtually filled with sand or small beads, typically around 1 mm in diameter. The beads are usually glass or ceramic, and the chamber is fitted with a screen to retain the beads within the chamber. As the shaft rotates, the dispersion is pumped through the mill and is subjected to considerable shear and impact forces from the violently agitated beads. Bead and sand mills are small, relatively quiet with a low power consumption, and can be used as a continuous process as well as a batch operation, giving increased flexibility. They do, however, need an efficient premixing operation, and can be difficult to clean, which may give problems with cross-contamination unless each type of dispersion has a dedicated mill. The small size of the beads used (typically 1 mm) can also give rise to screen blinding as the beads wear. It is not uncommon to see bead or sand mills used in conjunction with a ball mill or an attritor – the dispersion is initially milled by the ball mill and then passed through the bead mill for a final particle size reduction.

In many of these comminution processes the dispersion is pumped to circulate it through the mill. It is very important when pumping a dispersion not to allow the pump to stand idle for any length of time. Some dispersions can sediment very quickly and if solid materials are allowed to settle out in the pump chamber you will have a problem. No pump will work when the pump chamber is half full of sedimented sulfur.

2.4.5 High-shear mixers

High-shear mixers are powerful stirrers that rotate at very high speed. They can be fitted with impeller discs resembling a saw blade or with a head that can grind dispersions or pump fluids. Whilst they are not very efficient at breaking down the primary particles of a solid, they are very useful for emulsifying liquids, premixing dispersions and redispersing sedimented dispersions. The impeller head can usually be easily changed, making them a very versatile addition to a compounding department.

2.5 Measurement of particle size

The particle size of the dispersions can be measured in several ways. Electrical methods such as the Coulter counter, which monitors the change in electrical conductance as the latex or dispersion particles are drawn through a small aperture, are commonly used. The particle size can also be measured by light scattering, where the extent to which a coherent beam of light is scattered by the particles is measured and used to calculate the particle size and particle size distribution. Both of these methods rely on algorithms to calculate the particle size and various assumptions about the particle shape are made. Hence, the calculated particle 'size' will depend upon the parameters entered and may not be the actual physical size. The particle size may also be estimated by direct observation, e.g., using a microscope. A suitable graticule can be fitted to the microscope to give a direct measure of size, and the presence of Brownian motion is also a good indication that a small particle size has been achieved. Typical magnifications required for this type of observation will be around 125X. The ultimate method to determine particle size is with an electron microscope, although this will be beyond the resources of most dipping companies. There are also other, more subjective tests, such as observing the appearance and rate of sedimentation of a drop of the dispersion when dropped into water (sometimes called the 'cloud' test). For the cloud test, it is important to let the water stand undisturbed for a while to ensure that all turbulence has died away. Alternatively the Hegman, or grind gauge can be used, in which a diluted dispersion is spread into a finely machined groove of tapering depth using a straight-edged blade. The point at which the dispersion can no longer be accommodated in the groove indicates the minimum particle size. It will have been noted that all of the tests described above use a very small amount of dispersion as the sample and so it is very important to ensure that the dispersion is sufficiently agitated prior to sampling in order to ensure that the results are representative of the batch. Any of these methods, used carefully, will allow the particle size of the dispersion to be monitored with an acceptable degree of accuracy. However, it is recommended that the more subjective methods (e.g., the cloud test) are verified against more objective standards occasionally.

2.6 Storage of dispersions

Most of the dispersions of compounding ingredients, especially ZnO, sulfur and most accelerators, are heavier than water, and will thus tend to sediment. A list of the specific gravity of some of the more widely used materials is shown in Table 2.5.

Any sedimentation of the dispersion will reduce the concentration of active material in the bulk, and thus a procedure must be in place either to prevent this sedimentation, such as continuous agitation, or to redisperse the material to a satisfactory standard. A problem with continuous agitation of dispersions is that the dispersion can dry and cake on the stirrer shaft, and any splashed dispersion can also dry on the walls. This caked material can fall back into the dispersion to give large particles which, if allowed to go into the latex dipper, can cause problems with latex stability and give rise to film defects in the finished product. Allowing the dispersion to sediment, and only redispersing it when required also has its problems. Sulfur in particular gives a compacted sticky sediment, which can be difficult to break up again, and the sediment produced by a ZnO dispersion poses a similar problem. These sediments can be redispersed by manual mixing or mechanical mixing, but a final session with a high-shear mixer will generally be required to ensure complete redispersion and to give a homogeneous material. The problems inherent in the storage of dispersions are one of the reasons why it is recommended that the total solids concentration (TSC) of dispersions (as an indication of the concentration of active material) is checked before compounding. It is possible to slow down the rate of sedimentation and compaction *via* the dispersion formulation, for example, by incorporating viscosity modifiers to reduce the rate of sedimentation and by controlling pH. Examples of such formulations are given in Section 2.4. Due to the problems of dispersion storage, some companies elect to prepare only enough of the dispersions for one compounding batch at a time, thus avoiding the need for storage.

Table 2.5: Specific gravities of some materials commonly used in latex compounding.

Material	Specific gravity
ZnO	5.67
Sulfur	2.06
ZDEC	1.465
Zinc dibutyldithiocarbamate	1.27
Zinc dimethyldithiocarbamate	1.66
Zinc dibenzyldithiocarbamate	1.41
Zinc ethylphenyldithiocarbamate	1.50
Zinc diisononyldithiocarbamate	0.9
Zinc pentamethylenedithiocarbamate	1.60
Tetramethylthiuram disulfide	1.43
Tetramethylthiuram monosulfide	1.39
Zinc isopropyl xanthate	1.56
Zinc 2-mercaptobenzothiazole	1.69
2-Mercaptobenzothiazole	1.50
Tetraethylthiuram disulfide	1.31
Diphenylthiourea	1.32
Diphenylguanadine	1.13
Antioxidant 2246 [2,2′-methylene-*bis*-(4-methyl-6-*tert*-butyl phenol)]	1.08
Wingstay L	1.1
Antioxidant 425 [2,2′-methylene-*bis*-(4-ethyl-6-*tert*-butyl phenol)]	1.1
Butylated hydroxytoluene	1.05
Vanox 102	1.08

ZDEC: Zinc diethyldithiocarbamate

2.7 Compounding procedure

2.7.1 Representation of a latex formulation

There are specific conventions to be used in representing a rubber formulation, and hence also a latex formulation, which are designed to avoid confusion. For example, to add 5% of water to NR latex, the problem arises as to whether to add: a) 5 g of water for every 95 g of latex? b) 5 g of water for every 100 g of latex? or c) since the latex is 60% rubber, should we add 5 g of water to every 167 g of latex?

To avoid confusion arising from this type of question it is conventional to work in parts per hundred parts of rubber (phr). To work in this way, all calculations are based on 100 parts (usually by mass) of 'dry' rubber. So, for example, in a raw NR latex of 60% dry rubber content (drc), there are 60 parts of the rubber in 100 parts of the latex, or 100 parts of the rubber in 166.67 parts of the latex (this figure is achieved by the calculation 100/60 × 100). As it is usual to work in units of mass, we see that 166.67 kg of NR latex will contain 100 kg of NR. 167 g of the latex would contain 100 g of dry rubber. 50 g of the same latex would contain 30 g of the rubber (50/166.67 × 100) and so on. For a nitrile rubber (NBR) latex of 45% drc, 100 g will contain 45 g of the rubber, and 100 kg of this rubber will be found in 222.22 kg of the wet latex. Therefore, if we have an ingredient added at a level of 1 phr, there will be 1 g of that ingredient for every 100 g of rubber, or every 167 g (rounding to the nearest g) of NR latex at 60% drc or for every 222 g of NBR latex at 45% drc. It is most unusual to add ingredients to latex in a dry form – usually they are in the form of emulsions, solutions or dispersions, and so will be diluted in one way or another. In this case, we need to correct for the activity of the ingredient in the carrier. For example, if we want to add 1 g of KOH and the KOH is available as a 10% solution, we would need to add 1/10 × 100 or 10 g of the solution. Similarly to add 1 kg of sulfur dispersion at an activity of 50% we would need to add 1/50 × 100 or 2 kg of sulfur dispersion. So to answer the question posed above, we would first need to redefine the amount to be added as phr, not as a percentage. Then to add 5 phr of water would be answer 'c)', 5 g of water to 100 g of rubber, or 167 g of latex at 60% drc.

Latex formulations are usually represented in this way, and a worked example using a typical NR formulation is shown below. The first step is to assume an initial concentration of rubber as 100 parts (usually by mass) of dry rubber in an appropriate template (Figure 2.3).

Batch size (kg):				**Scale factor :**			
Material	**Activity (%)**	**Formulation**		**TSC (%)**			**Amount to add**
		Dry	**Wet**	**Theory**	**Actual**	**Correction**	
NR latex		100					

Figure 2.3: Template for a rubber latex formulation.

Then, we add the activity of the latex – in this case the drc (60%). This allows us to calculate the amount of latex that will contain 100 parts of dry rubber as 100/60 × 100 = 166.67 to two decimal places. At the same time we can insert the batch size for which we want to calculate the formulation. Let us assume that this latex compound is for laboratory dipping and around 11 kg of compounded latex will allow a full evaluation, so we will base this batch on 10 kg of raw latex. Putting this into the template gives us Figure 2.4.

Now we can add the rest of the compounding ingredients and the relevant activities (Figure 2.5) using, for this example, a simple NR latex formulation. Note that the activities of the various solutions and dispersions are usually simple proportions to aid in subsequent calculations.

Batch size: 10 kg raw latex				Scale factor:			
Material	Activity (%)	Formulation		TSC (%)			Amount to add
		Dry	Wet	Theory	Actual	Correction	
NR latex	60	100	166.67				10.00 kg

Figure 2.4: Partially completed template for a rubber latex formulation.

Batch size: 10 kg raw latex				Scale factor:			
Material	Activity (%)	Formulation		TSC (%)			Amount to add
		Dry	Wet	Theory	Actual	Correction	
NR latex	60	100	166.67				10.00 kg
KOH	10	0.20					
Potassium oleate	20	0.25					
Antioxidant	50	0.50					
Sulfur	50	0.75					
ZDEC	33.3	0.55					
ZnO	50	0.45					
Water	100	20.00					

Figure 2.5: Partially completed template for a rubber latex formulation.

We convert the dry proportion to the proportions to be actually used, that is as a solution, dispersion and so on – the 'wet' proportion – by dividing the dry proportions by the activity and multiplying by 100 (Figure 2.6).

We can now calculate the amount of each material to add. We have decided to use a quantity of 10 kg of raw latex. To convert 166.67 parts of the latex, as shown in the formulation, to 10 kg we need to divide 10 by 166.67 = 0.06. This then becomes the scale factor and we similarly convert all the wet quantities to the actual quantity to add by multiplying them by this scale factor, as shown below (Figure 2.7). The number of significant figures to be used will, to some extent, depend on the application and the materials in question. For example, it would be unnecessary (and impractical) to weigh out 1,000 kg of latex to an accuracy of ±10 g, whereas with a non-ionic surfactant, where low concentrations are generally used and where small amounts can have a big effect, it would be reasonable to specify a tolerance of ±0.1% on the amount to be added. The best advice is to be sensible and adopt a practical approach to the tolerances in the latex formulations.

Batch size: 10 kg raw latex				**Scale factor:**			
Material	**Activity (%)**	**Formulation**		**TSC (%)**			**Amount to add**
		Dry	Wet	**Theory**	**Actual**	**Correction**	
NR latex	60	100	166.67				10.00 kg
KOH	10	0.20	2.00				
Potassium oleate	20	0.25	1.25				
Antioxidant	50	0.50	1.00				
Sulfur	50	0.75	1.50				
ZDEC	33.3	0.55	1.65				
ZnO	50	0.45	0.90				
Water	100	20.00	20.00				

Figure 2.6: Partially completed template for a rubber latex formulation.

Batch size: 10 kg raw latex				**Scale factor:** 0.06			
Material	**Activity (%)**	**Formulation**		**TSC (%)**			**Amount to add**
		Dry	**Wet**	**Theory**	**Actual**	**Correction**	
NR latex	60	100	166.67				10.00 kg
KOH	10	0.20	2.00				0.12 kg
Potassium oleate	20	0.25	1.25				0.075 kg
Antioxidant	50	0.50	1.00				0.06 kg
Sulfur	50	0.75	1.50				0.09 kg
ZDEC	33.3	0.55	1.65				0.099 kg
ZnO	50	0.45	0.90				0.054 kg
Water	100	20.00	20.00				1.20 kg

Figure 2.7: Example of a rubber latex formulation.

This gives us a formulation containing actual amounts that can be weighed out and used in the laboratory to compound this trial batch of latex. However, it is necessary to recognise that the TSC of dispersions can and does change through compounding variations, sedimentation or evaporation of liquid. Experience has shown that the TSC of solutions is less likely to change, although it is not unknown. For this reason it is good practice to measure the total solids of the compounding dispersions at intervals. If the values of TSC have moved away from the specification value a correction factor can be applied, and this is the reason for the hitherto unused columns in the formulation sheet. The correction factor is conveniently based on the TSC – it is much easier to measure total solids than activity. The assumption is made that if the TSC has changed, the activity has changed by the same amount. So we can complete the example of a formulation sheet (Figure 2.8) by including the TSC results for the dispersions.

Batch size: 10 kg raw latex				**Scale factor:** 0.06			
Material	**Activity (%)**	**Formulation**		**TSC (%)**			**Amount to add**
		Dry	**Wet**	**Theory**	**Actual**	**Correction**	
NR latex	60	100	166.67	–	–	–	10.00 kg
KOH	10	0.20	2.00	–	–	–	0.120 kg
Potassium oleate	20	0.25	1.25	–	–	–	0.075 kg
Antioxidant	50	0.50	1.00	51.3	51.55	0.995	0.060 kg
Sulfur	50	0.75	1.50	53.4	53.10	1.006	0.091 kg
ZDEC	33.3	0.55	1.65	35.2	34.95	1.007	0.100 kg
ZnO	50	0.45	0.90	50.8	51.66	0.983	0.053 kg
Water	20	100	20.00	–	–	–	1.200 kg

Figure 2.8: Example of a rubber latex formulation.

It will be noted that in this example the changes resulting from the deviations of the dispersions from the specification values are small, and in practice this will usually be the case. It will be up to the individual factory and compounding department to decide whether to apply a correction or not. Limits can be set for the deviation of TSC, beyond which a correction factor must be applied. For example, the TSC for the antioxidant in the above example could be 51.3 ± 0.25. If the results lie inside the range 51.05 to 51.55, then no correction would be necessary.

2.7.2 Addition of compounding ingredients

The latex and other compounding ingredients should be weighed or measured out to a sensible degree of accuracy, bearing in mind the material and the effect it will have on the finished compound. Non-ionic surfactants, for example, which are added at low concentrations and have a powerful effect, must be weighed out very accurately. In general, an accuracy of ±0.1% should prove to be entirely satisfactory. The balances and graduated containers must be able to achieve the required degree of accuracy and precision.

The latex should be transferred to the compounding tank in a manner that will not cause problems with coagulum or air entrainment. Transfer by hand is adequate for small quantities, but for larger quantities it is necessary to use a suitable pump (e.g., a diaphragm pump) or, even better, store the latex at a higher level, so that it can be transferred to subsequent stages within the compounding cycle by gravity. The latex is preferably filtered when it is added, to remove any skin or coagulum that may have built up during storage. It is also advisable to try and avoid the generation of foam, as the foam may dry out to give particles of rubber that could cause problems in later stages of the manufacturing.

Once the latex is added to the compounding tank, start the agitation and add the compounding ingredients. There are two rules to be followed here: add the stabilising materials first and do not add them too quickly. A third rule could be added: filter all the materials before, or on, addition. Some of the common compounding materials – those containing zinc for example – will destabilise latices, and it is necessary to protect the latex as much as possible before these are added. This is the reason materials that stabilise the latex, such as surfactants and alkalis, are added first. Don't forget that water can also function as a stabiliser, as dilution will reduce the frequency of interparticle collisions, but make sure that sufficient water is held back to dilute the compounding ingredients as necessary and rinse out the containers. After the stabilising ingredients are added, materials which have no or little effect on stability, such as dispersions of sulfur and antioxidants, can be added. Finally, add those components that have the capability to destabilise latex. These can include zinc-containing materials such as ZnO or zinc-based accelerators, and acidic materials such as mercaptobenzothiazole. It is also recommended to try and match, as far as possible, the pH of any additives with that of the latex, by, for example, adding a small amount of ammonium hydroxide (NH_4OH) or KOH to the emulsion or dispersion. Be careful, however, with carboxylated latices, such as many of the NBR latices as an incautious addition of alkali such as KOH can cause the latex to thicken. Dilute the ingredients before addition using, for example, water containing a small amount (say 1–2%) of NH_4OH. This will reduce the ionic strength of the materials and thus reduce the possibility of ionic shock to the latex. As a guideline, try to dilute the materials at least 50%. The compounding ingredients should be added slowly, again to avoid or reduce shock to the latex. It is suggested that an addition rate of no faster that 1 or 2 litres a minute is used, although this will depend upon individual circumstances.

Note that it is very important that the dispersions are agitated well before use and the pH checked before addition to maximise compatibility with the latex. As mentioned earlier, the TSC of a dispersion can drift with time, owing to sedimentation and the evaporation of water. It is therefore common practice to check the TSC at intervals or before use, and use this concentration as a guide to the activity of the dispersion (it is much easier to measure TSC than activity, and it is a reasonable assumption that if the TSC is within specification, so is the concentration of the active ingredients). If the TSC has drifted out of specification, the amount of dispersion to add should be modified to compensate. As recommended above, dispersions should be filtered through a fine Nylon or stainless steel mesh to remove large and agglomerated particles before use. Such particles in the latex will sediment more easily, may initiate some coagulation (especially any zinc-containing materials) or instability in the latex and will, if incorporated into the film, create a potential flaw. If undamaged, the filter mesh can be rinsed out and reused time and again.

2.7.3 Latex prevulcanisation

Once all the ingredients have been added to the latex it will be either left to mature or prevulcanised. It is most unusual for a latex to be dipped immediately after compounding. NR latex and synthetic *cis*-1,4-polyisoprene rubber (IR) latex will usually be prevulcanised, whereas CR and NBR latices are more often allowed to mature at room temperature (RT) for a period of time. Prevulcanisation temperatures of up to 80 °C have been used with NR latex, although generally much lower temperatures – say 40 to 60 °C – will be employed. With NR latex the physical properties of the finished product will generally be better with lower prevulcanisation temperatures, although of course the duration of the prevulcanisation will need to be longer. In these cases, a compromise between acceptable time and properties will need to be established. If RT prevulcanisation or maturation is required, the latex can now be left for the specified time, although it is recommended that the agitation is continued for a while. If the latex is to be left without agitation the surface must be covered to prevent a skin being formed when water evaporates from the surface, and thin polyethylene sheeting is ideal for this purpose. If NR and IR latex are left standing for any length of time, even when covered, they will cream, as the lighter rubber particles rise to the surface. Any residual solid particles of dispersion will also fall to the bottom; hence, before use, the latex must be agitated again. CR and NBR latices are much less prone to creaming, but will still need to be agitated before use to ensure homogeneity.

The maturation time will depend on the latex, the formulation, the application and the production requirements, and can be anything from a few hours to a few weeks. In some cases a second compounding and heating stage, following the guidelines above, may be employed before the latex is used.

A final compounding stage is carried out when the maturation is deemed to be complete. The latex will be adjusted to the required viscosity for dipping by the addition of water, and any other additions, such as pigment, will be made.

When a latex is prevulcanised, crosslinks are created within the individual rubber particles. When done correctly, the latex will maintain its stability and character. Indeed, in the case of NR and IR the crosslinking reactions proceed much more easily and rapidly in the latex state than in the dry state [5], as can be seen in

Figure 2.9. The graph shows the rate of increase of network-combined sulfur, at 50 °C, for the same NR formulation. In both cases the formulation was compounded as a latex, but in one case the latex was air-dried at RT before being heated. As can be seen, the sample remaining as a liquid prevulcanises much more rapidly at 50 °C than the same formulation when dried. The only difference between the two is the presence of water.

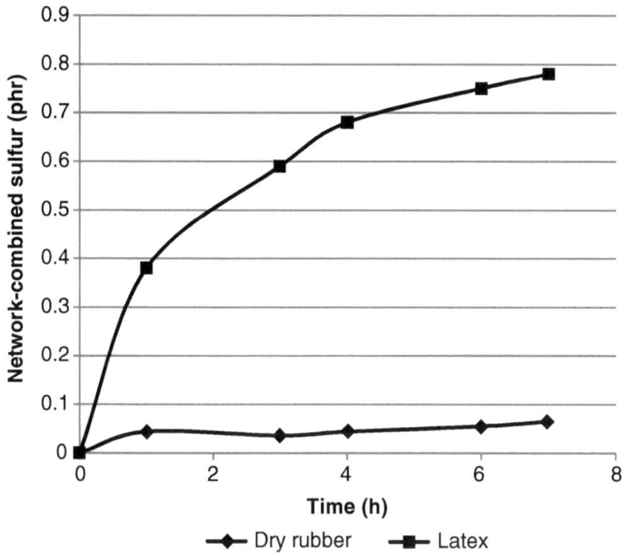

Figure 2.9: Difference in the rate of prevulcanisation between compounded NR as a latex and as an air-dried film. Adapted from A. C-P. Loh in *Further Investigations of the Prevulcanisation of Natural Rubber Latex*, The North London Polytechnic, London, UK, 1982 [PhD Thesis] [5].

It is worth noting at this point that latices can also be prevulcanised by peroxides and ionising radiation, although at the time of writing these procedures are not widely used in the latex dipping industry and will not be discussed further.

This easy prevulcanisation has several implications for cost and quality. The prevulcanisation reaction can proceed at a low temperature – with suitable accelerators, prevulcanisation will take place at temperatures of 30 °C or lower, which is at RT in many latex manufacturing countries. Using a low temperature means that latex stability is less likely to be compromised. The latex can also be prevulcanised in bulk. Heating the latex in a large insulated tank allows more control over critical aspects such as temperature, maintenance of pH, control of solids content and so on, and is much more energy efficient than heating as a dipped film, where the not inconsiderable mass of a former and former mounting also have to be heated and, as shown above, a dry film is very much more difficult to crosslink. As well as the energy cost of vulcanising a thin, dried film, the extra heat required means that the risk of oxidation and degradation will be much higher.

The physical properties of products dipped from a prevulcanised latex can be affected by the extent of prevulcanisation. As the prevulcanisation reactions proceed, the latex particles become harder, less ‘sticky’, and the particles in a highly prevulcanised latex will not integrate as completely within the finished film. The physical properties and resistance to solvents of the resulting film will be reduced. On the other hand, this incomplete particle integration means that a prevulcanised latex will dry faster, offering more possibilities for energy saving, and film integration can be enhanced subsequently by hot water leaching. In practice, NR films with an entirely satisfactory level of physical properties can be obtained from prevulcanised latices. As a prevulcanised latex will not require much postvulcanisation after dipping – sometimes only drying – the level of curatives in the dip tank will be lower, giving less sedimentation of solid materials, greater colloidal stability, greater cure stability of the latex and more consistency of performance. All in all, it is better to prevulcanise those latices that can be prevulcanised, and to do so to the maximum extent consistent with acceptable properties of the finished product.

To prepare a prevulcanised latex in practice, the compounding will generally be carried out as usual. The latex will then be heated, with agitation, to the required prevulcanisation temperature and held there for the required time. It is strongly recommended that agitation is continued during prevulcanisation. The prevulcanisation time will generally be short – a few hours at high temperatures and rather longer at lower temperatures, depending upon the cure system in use. As stated earlier, RT prevulcanisation is possible in tropical countries, albeit at longer prevulcanisation times. After the required time the latex will be cooled. A period of storage, generally without agitation, will allow unwanted curatives to sediment out and the latex to mature, or additional curatives can be added if it is desired to promote further reaction. The actual reactions that occur during maturation are not well understood, but in the author’s opinion, a prevulcanised latex will always be better after maturation.

If prevulcanisation is to be carried out at a high temperature it may be best not to use an ethoxylated non-ionic surfactant as part of the stabiliser system, as the solubility of these materials can be reduced at elevated temperature. If it is intended to use a fully prevulcanised system, so that no additional curing is needed (also known as a ‘dip and dry’ system), then no curatives will be needed in the dip tank. As a consequence, it is not necessary to grind the compounding ingredients to such a fine particle size, so that any residual curatives can sediment out more easily. As pointed out earlier, this larger particle size will not affect the rate of reaction.

The mechanism of sulfur prevulcanisation is presumed to follow a similar path to that of accelerated sulfur vulcanisation described in Chapter 3, with the added complication that it is necessary to transfer the curatives from their dispersions to the rubber particle. Surprisingly, as we have seen above, the reaction is faster than with intimate mixing of the dry rubber sample. It has been shown that sulfur is very rapidly transported into the rubber [5], whilst the transport of a zinc dialkyldithiocarbamate accelerator, in the absence of sulfur, is very slow. The accelerator is transported more quickly when sulfur is present, leading to the presumption that the transport is in the form of an activated sulfur/accelerator complex. At the risk of oversimplifying, the rate of prevulcanisation tends to be zero-order with respect to accelerator concentration – possibly because of the difficulty in transporting the material into the rubber particle – and first-order with respect to sulfur. In other words, the sulfur concentration is more likely to affect the reaction rate than the accelerator concentration. If you want the reaction to proceed more rapidly, add more sulfur and *vice versa*.

2.7.4 Measurement of prevulcanisation

The compounded latex will generally be monitored throughout the compounding stage to ensure that it meets the requirements as defined in any company procedures. Typical parameters will include viscosity, pH, total solids content and possibly mechanical stability. One of the main parameters for natural and synthetic polyisoprene latices will be the state of any prevulcanisation or maturation. Several test methods exist, of differing speed, accuracy and utility. With NR, the easiest and fastest is the 'chloroform number' where a small sample of latex (say 10 ml) is stirred with an approximately equal volume of chloroform. The chloroform is absorbed into the rubber particles and a coagulated mass ensues. The degree of prevulcanisation is estimated by the appearance of the coagulum and given a number from 1 to 4. Chloroform number 1 equates to a low degree of precure and 4 to a high degree of precure. The description and assessment of the different numbers is not really standardised, but typically can be described as shown in Table 2.6.

Table 2.6: Grades in the chloroform test for prevulcanised NR.

Chloroform number	Appearance of the coagulated mass
1	Soft and sticky. When pulled apart will form a string before it breaks
2	Soft, only slightly sticky and breaks short when pulled apart
3	The coagulated latex is in the form of large lumps, which are not tacky
4	The coagulated latex is in the form of small dry crumbs

Sometimes intermediate numbers are used, for example 2.5 and 3.5, although the reproducibility of these is questionable. Although quick and easy, the chloroform number method is rather subjective and does not discriminate well at higher levels of cure. A more accurate method is to swell a dried sample of the material in a suitable solvent. For speed, the sample is usually a thin film drawn down from the latex, which can be dried quickly at room temperature (positioning in front of a fan will speed up the drying process). A thin and even film from latex is readily obtained using one of the commercially available bar applicators or draw-bars, on, for example, a clean glass plate or specially treated paper. After drying, the sample is powdered (to remove tack) and carefully peeled off the substrate. A sample of known dimensions or weight is cut from the film and immersed in a suitable solvent. The film sample will swell until an equilibrium swelling is reached, dictated by the number of crosslinks in the film. This extent of swelling can be assessed either by mass (or volume fraction of rubber or solvent in the swollen test piece) or size (linear or area swell). The results have an inverse relationship – the greater the extent of crosslinking, the lower the swell – and are more accurate with lightly crosslinked samples than heavily crosslinked ones. Suitable solvents are toluene for NR and a more polar solvent, such as methyl ethyl ketone, for NBR. Finally, the degree of prevulcanisation can be measured by physical measurements. The modulus of a test sample of rubber is directly related to the number of crosslinks in the sample, and so a tensile or relaxed modulus test will indicate the degree of prevulcanisation. The most rapid of these measurements is the prevulcanisate relaxed modulus test [6], in which a tube is dipped into the latex and withdrawn. The coating of latex is quickly dried (a few minutes at 70 °C is usually sufficient), rolled off the tube, weighed and the modulus of the ring after 1 min at 100% extension is measured. All of these methods have their drawbacks. As mentioned above, the chloroform method is subjective and not sensitive at higher crosslink densities, such as is commonly used with prevulcanised NR latex. Swelling tests are more objective, but are also less responsive at higher crosslink densities (although better than the chloroform test). Modulus tests are the most accurate, but require more sample preparation and test equipment. Gorton [7] compared different methods of measuring precure on a compounded NR latex after various heating times and the results are plotted in Figures 2.10–2.13. As can be seen, the chloroform method (Figure 2.10) does not detect any change beyond 4 h heating, and the results before then are not very progressive. The swelling method (Figure 2.11) similarly becomes less sensitive as the prevulcanisation reaction proceeds, although it is better than the chloroform test. Both the tensile modulus (Figure 2.12) and relaxed modulus (Figure 2.13) methods show a more progressive response and can discriminate between the rubber samples as the heating period, and presumed crosslink density, progresses.

Other quality control tests that will usually be applied to compounded latex include pH, viscosity, total solids content, ammonia content and sometimes a test of stability such as mechanical stability time or one of the zinc stability tests.

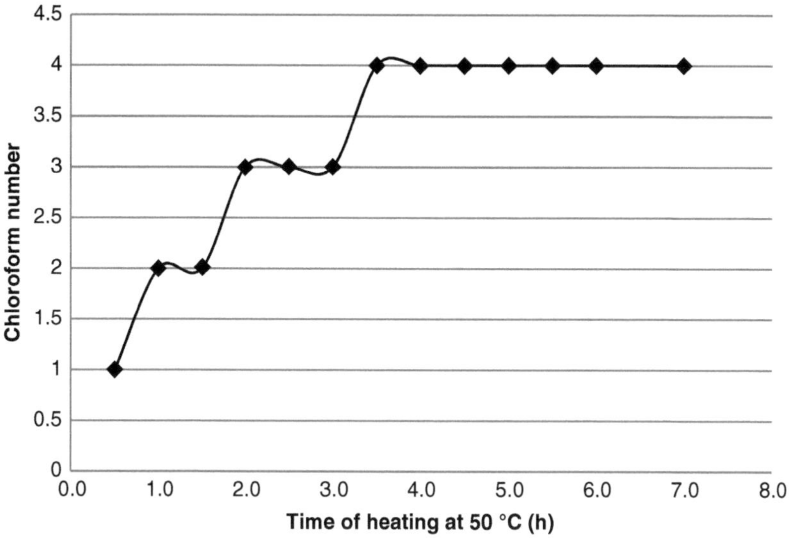

Figure 2.10: Chloroform number of a compounded NR latex with time of heating at 50 °C. Reproduced with permission from A.D.T. Gorton, *NR Technology*, 1979, **10**, Part 1. ©1979, Tun Abdul Razak Research Centre, UK [7].

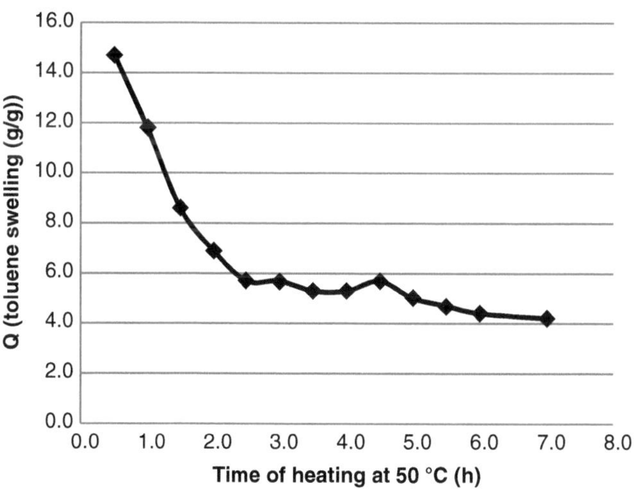

Figure 2.11: Swelling in toluene of a film from a compounded NR latex against time of heating at 50 °C. Reproduced with permission from A.D.T. Gorton, *NR Technology*, 1979, **10**, Part 1. ©1979, Tun Abdul Razak Research Centre, UK [7].

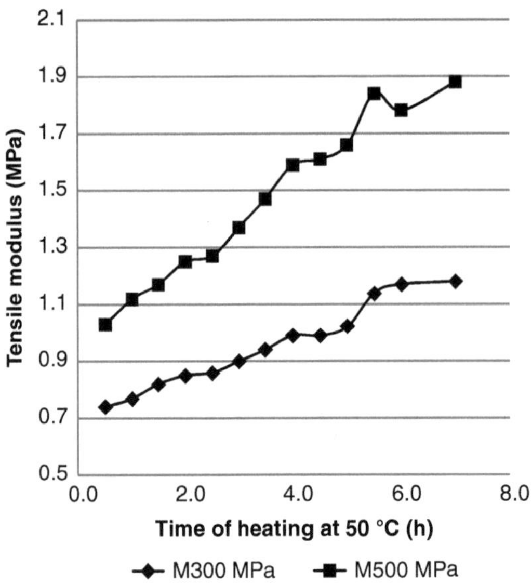

Figure 2.12: Tensile modulus of a film from a compounded NR latex against time of heating at 50 °C. Reproduced with permission from A.D.T. Gorton, *NR Technology*, 1979, **10**, Part 1. ©1979, Tun Abdul Razak Research Centre, UK [7].

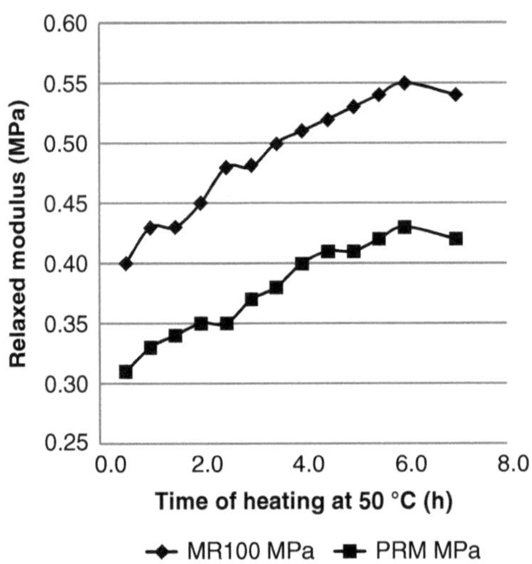

Figure 2.13: Relaxed modulus of a film from a compounded NR latex against time of heating at 50 °C. Reproduced with permission from A.D.T. Gorton, *NR Technology*, 1979, **10**, Part 1. ©1979, Tun Abdul Razak Research Centre, UK [7].

2.8 Storage of compounded latex

It is possible to store prevulcanised compounded latex for some time, provided that certain precautions are taken. The formulation must be cure stable – meaning that there is little or no residual sulfur – so prevulcanisation does not continue during storage. The latex should be stored in sealed containers, clearly labelled so that it cannot be used mistakenly. Storage temperatures should be moderate – ideally between 10 and 30 °C. There should be a documented and validated testing procedure to confirm that the latex remains fit for purpose, and the latex must comply with the requirements of this procedure before it is used.

If these conditions are met, then compounded prevulcanised latex can be stored for several months, although it is recommended that any stored latex is blended with fresh prevulcanised latex before use.

References

1. A.D.T. Gorton and T.D. Pendle, *NR Technology*, 1981, **12**, 2, 21.
2. M. Porter, R. Rawi and S.A. Rahim, *Journal of Natural Rubber Research*, 1992, **7**, 85.
3. C. Petri in *Proceedings of Latex 2004*, Hamburg, Germany, 2004, Paper No.3.
4. T.C. Patton in *Paint Flow and Pigment Dispersion*, Wiley Interscience, Hoboken, NJ, USA, 1979.
5. A. C-P. Loh in *Further Investigations of the Prevulcanization of Natural Rubber Latex*, The North London Polytechnic, London, UK, 1982 [PhD Thesis].
6. A.D.T. Gorton and T.D. Pendle, *NR Technology*, 1976, **7**, 77.
7. A.D.T. Gorton, *NR Technology*, 1979, **10**, Part 1.

3 Vulcanisation

3.1 Introduction

Vulcanisation is critical to all rubber articles, except those made from thermoplastic elastomers. The chemistry behind the sulfur vulcanisation of elastomers is complicated, hence only a very brief and simplified description will be given here. For more information the interested reader is referred to the comprehensive review by Chapman and Porter [1]. This review focuses on natural rubber (NR), predominately dry NR, but it is thought that much of the mechanism described can apply to the sulfur vulcanisation of other diene-containing elastomers.

Vulcanisation, also known as crosslinking or curing, is the process whereby the elastomer chains are joined together, increasing strength and elasticity, and reducing the tendency of the rubber to creep when extended. On a molecular level, elastomers exist as single polymer chains. A certain degree of strength and elasticity is present, through entanglement of the chains, but this strength and elasticity is not high. Vulcanisation introduces crosslinks at various points to tie the separate polymer chains together, which can greatly enhance strength and elasticity. Table 3.1 shows typical results for an unfilled NR compound, before and after sulfur vulcanisation. As can be seen, both the tensile strength and tensile modulus are greatly increased, whilst the elongation is only slightly reduced.

Vulcanisation is named after Vulcan, the Roman god of fire, as the original vulcanisation processes in the 19th century involved mixing the rubber with sulfur and heating for a long time. Although this introduced crosslinks into the rubber, it also degraded it to a considerable extent, so the finished properties were not very good. A big step forward in vulcanisation was the discovery of rubber accelerators in the early 20th century, initially based on compounds derived from aniline. These organic accelerators greatly increased the speed with which sulfur could be combined with rubbers, opening up not just the rubber industry to greater development, but also starting the rubber accelerator industry which now provides hundreds of compounds, some specifically tailored to particular applications.

3.2 Brief overview of the mechanism of sulfur vulcanisation

So how does sulfur vulcanisation of a latex elastomer occur? As mentioned earlier, the picture is complex and not yet fully understood. NR is the most studied elastomer, and current thinking proposes the following mechanism. The initial step is the formation of an 'active sulfurating agent', formed by the reaction of

https://doi.org/10.1515/9783110638097-003

Table 3.1: Typical results for an unfilled NR compound, before and after sulfur vulcanisation.

Property	Typical results for NR	
	Unvulcanised	Vulcanised
Tensile strength (MPa)	8	30
Modulus (S_{300}, MPa)	0.5	1.3
EB (%)	920	850

EB: Elongation at break

sulfur (present as the S_8 molecule) with the accelerator to create a perthiomercaptide (Equation (3.1)):

$$\begin{array}{ccc} XS-Zn-SX+S_8 & \rightarrow & XS-S_a-Zn-S_b-SX \\ \text{Accelerator entity} & & \text{Perthiomercaptide} \end{array} \tag{3.1}$$

This reaction is facilitated by the presence of ammonia, or other amines, which act as ligands to create a small electrical dipole on the accelerator molecule, thus activating the zinc–sulfur bonds in the accelerator. This active sulfurating agent can then react with a rubber molecule to insert a short sulfur chain into the polymer chain, forming what is called a 'rubber-bound intermediate' (Figure 3.1).

$R\text{-}S_y\text{-}X$

Figure 3.1: General structure of a rubber-bound intermediate. 'R' is the rubber molecule, 'S_y' is a short polysulfidic chain and 'X' is an entity derived from the accelerator.

This rubber-bound intermediate can then react with other rubber molecules or with other rubber-bound intermediates to form the initial crosslinked network (Figure 3.2).

$R\text{-}S_y\text{-}X + R\text{-}H \rightarrow R\text{-}S_x\text{-}R$

$2\ R\text{-}S_y\text{-}X \rightarrow R\text{-}S_x\text{-}R + X\text{-}S_z\text{-}X$

Figure 3.2: Reactions of the rubber-bound intermediate to form the initial crosslinked network. 'R-H' represents the rubber hydrocarbon and S_x, S_y and S_z are short polysulfidic chains.

Both of these reactions are accelerated by zinc compounds, especially zinc accelerator-thiolate compounds.

At the same time, the residues from the accelerator can react with zinc oxide (ZnO) to regenerate the accelerator, which can then continue the reactions. We are not at the end of the process yet, however. This initial crosslinked network continues to react in various ways. The sulfur chains can break and reform, leading to crosslink shortening, lengthening or additional crosslinks being formed. The main polymer chain can be broken or modified, there can be sulfur–sulfur bond interchange, cyclisation and many other reactions. These will all be influenced by the vulcanisation conditions, temperature, vulcanisation components and their concentrations. The crosslinked network thus changes considerably with time, and this can be seen by the way the properties of the dipped products change over the first few days and weeks. A representation of the complexity of the rubber network is shown in Figure 3.3.

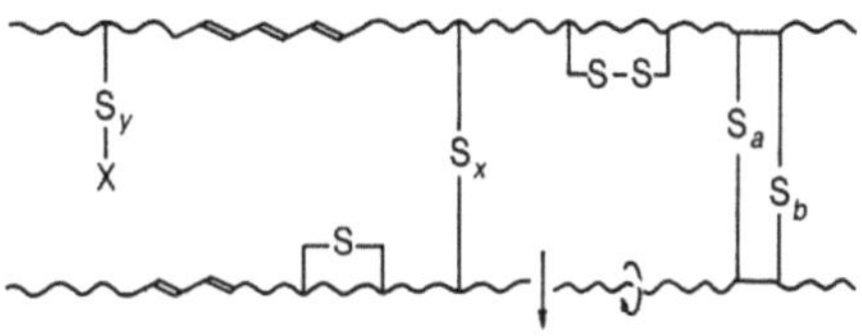

Figure 3.3: A representation of the complexity of a crosslinked rubber network. Reproduced with permission from A.V. Chapman and M. Porter in *Natural Rubber Science and Technology*, Ed., A.D. Roberts, Oxford University Press, Oxford, UK, 1988. ©1988, Oxford University Press [1].

The network will be changed again by ageing, adding to the complexity of the whole process.

In summary, the reaction pathway can be shown thus:

- Stage 1 – the vulcanising agents (accelerator, sulfur, activator) react to create the active sulfurating agent.
- Stage 2 – this active sulfurating agent reacts with the rubber molecules to give rubber-bound intermediates.
- Stage 3 – these rubber-bound intermediates can react with other rubber molecules to form the initial crosslinked network and/or two rubber-bound intermediates can add together, again forming the initial crosslinked network.
- Stage 4 – this is followed by assorted network maturation reactions, including:
  - Formation of additional sulfur crosslinks.
  - Destruction of existing sulfur crosslinks.
  - Modification of existing sulfur crosslinks.
  - Formation of cyclic sulfides.
  - Breakage of the main polymer chain.
  - Modification of the main polymer chain.
  - Sulfur–sulfur bond interchange.
- Stage 5 – over a period of time this will lead to the formation of the mature crosslinked polymer network.

As can be seen, the picture is very complicated.

This reaction scheme has been derived for dry NR, but it is thought that the picture is broadly similar for NR latex, with the additional complication that the reaction ingredients have to transfer through the aqueous phase to the rubber molecules. The aqueous phase can also contain materials such as ammonia and proteins that can act as ligands and thus assist in the vulcanisation reactions, making it both faster and more efficient.

With dry rubber one can encounter so-called efficient vulcanisation (EV) and semi-efficient vulcanisation (semi-EV) conditions. These are characterised by having a high accelerator concentration and a low sulfur concentration. In dry rubber, these conditions lead to a network having a high proportion of crosslinks containing only one (monosulfidic) or two (disulfidic) sulfur atoms. Networks of this type exhibit much better ageing resistance than those where the crosslinks contain a greater number of sulfur atoms (polysulfidic crosslinks). However, the conditions required to create this type of network require a high vulcanising temperature, whereas latex articles are usually vulcanised at much lower temperatures, as they are generally thin and have a high surface-to-volume ratio. They are thus more prone to oxidation at high temperatures, hence the use of active accelerators and low temperatures. As a consequence, dipped latex articles will have a network structure that is predominantly polysulfidic. This type of network will give good physical properties, but the ageing resistance may be less than for an EV or semi-EV system. However, as the NR chains in a latex system will not have been subject to mastication during compounding, unlike dry rubber, the polymer network will be stronger. This means that, in general, latex dipped NR articles will have better physical properties than those prepared from dry rubber.

3.3 Other vulcanisation procedures

Whilst sulfur vulcanisation is the major mechanism by which most elastomers are crosslinked, it is not the only one. Elastomers containing carboxylic groups, such as carboxylated nitrile rubber (XNBR) latex, can also be crosslinked ionically through the carboxyl group by metal oxides or alkalis. ZnO has been found to be soluble in XNBR [2], and it is thought that the reaction of XNBR with ZnO leads to the Zn^{2+} ion forming an ionic link between ionised carboxyl COO^- groups. These ionic crosslinks are very strong, but can be broken and reformed. Thus, articles crosslinked in this way will have high strength and modulus, but can also be prone to creep. Ionic crosslinks can also be disturbed by water. Similar crosslinking reactions have been found to occur with potassium hydroxide, which is thought to proceed through the formation of an ionomer system using the copolymerised carboxyl group [2]. The presence of olefinic unsaturation in the polymer backbone means that nitrile rubber (NBR) can also be crosslinked using a standard accelerated sulfur cure and frequently a mixed vulcanisation system will be used with these elastomers to fine-tune the physical properties. XNBR elastomers vulcanised with both sulfur and ionic bonds will show better durability in use than those crosslinked with ionic bonds alone. Carboxyl groups can also react with materials such as urea formaldehyde and melamine formaldehyde resins, but although these will crosslink XNBR they are not commonly encountered in the latex dipping industry.

Polychloroprene (CR) can also be crosslinked with metal oxides, this time reacting through one of the allylic chlorine atoms on the polymer chain. An allylic chlorine atom is one that is attached to a carbon atom, which itself is adjacent to a double bond. The electron resonance opportunities arising from this structure means that reaction intermediates are unusually stable, which in turn means that an allylic chlorine atom is very reactive. Chloroprene monomer polymerisation predominantly proceeds through 1,4-addition (Figure 3.4), but there will be a small proportion of 1,2-addition (Figure 3.5), giving rise to an allylic chlorine atom. The concentration of these allylic groupings in the CR chain is affected by the polymerisation conditions, but will usually be in the region of 1 to 2%.

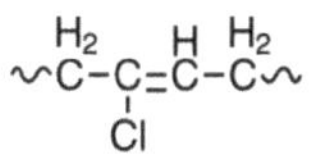

Figure 3.4: Structure resulting from the 1,4-polymerisation of chloroprene.

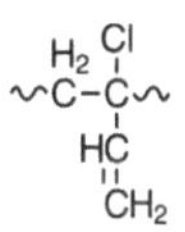

Figure 3.5: Structure resulting from the 1,2-polymerisation of chloroprene, creating an allylic chlorine atom.

Although there has been discussion about the mechanism by which CR is crosslinked, it is generally accepted that it proceeds through the *bis*-alkylation reaction of this active allylic chlorine atom [3]. There is also some olefinic unsaturation in the CR chain, allowing CR polymers to be crosslinked through an accelerated sulfur cure, although the electron-withdrawing properties of the chlorine atoms in the molecule mean that the polymer is much less active in this respect compared with NR or NBR polymers. Virtually all of the cure formulations for CR will use accelerators such as a thiourea, for example, diphenylthiourea (Figure 3.6), or a mixture of

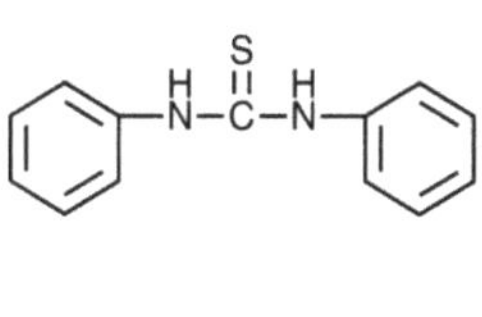

Figure 3.6: Diphenylthiourea.

thiourea and guanidine, such as diphenylguanidine (Figure 3.7). If a sulfur cure system is used as well, then other accelerators, such as a dithiocarbamate, may also be used.

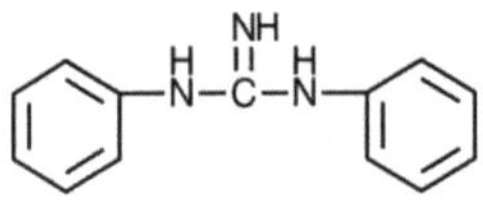

Figure 3.7: Diphenylguanidine.

Crosslinking of some elastomers can also be carried out by peroxides or radiation that will give carbon–carbon crosslinks. However, at present, these methods are of little significance in the industry.

3.4 Effect of vulcanisation on physical properties

The extent of vulcanisation can have various effects on a rubber product. Generally, increasing the crosslink density will increase the modulus and decrease the elongation at break. Both tensile strength and tear strength pass through a maximum as the crosslinking is increased, initially increasing and then decreasing. Typically, the crosslink density that gives the maximum value for these two parameters is different – one can either compound and vulcanise the latex to give the maximum tensile strength or the maximum tear strength, but not usually both. The typical response of some physical properties with increasing crosslink density is illustrated in Figure 3.8.

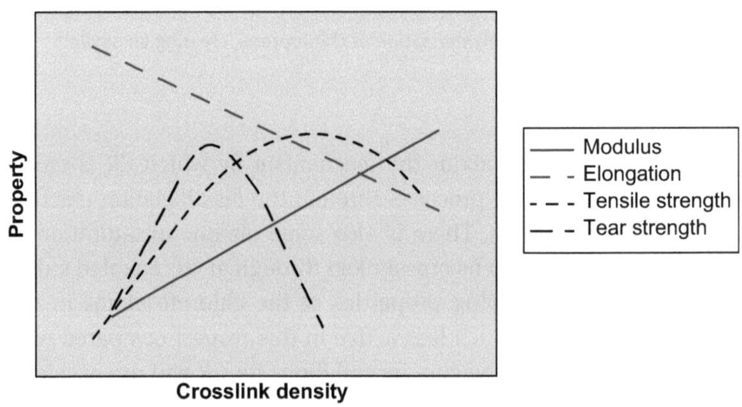

Figure 3.8: A typical response of some physical properties with increasing crosslink density.

As is frequently encountered when working with latex systems, a satisfactory compromise between conflicting parameters has to be found.

Note that in a sulfur-vulcanised system it is usually the amount of sulfur in the formulation that controls the rate and extent of crosslinking. As was pointed out earlier, the accelerator is regenerated by reaction with ZnO, so vulcanisation is not usually controlled by exhaustion of the accelerator provided that ZnO is present.

References

1. A.V. Chapman and M. Porter in *Natural Rubber Science and Technology*, Ed., A.D. Roberts, Oxford University Press, Oxford, UK, 1988.
2. A. Kells and R. Groves in *Proceedings of Latex 2006*, Frankfurt, Germany, 2006, Paper No.16.
3. P. Kovacic, *Industrial and Engineering Chemistry*, 1955, **7**, 5, 1090.

4 Degradation of elastomers

4.1 Introduction

Like many elastomers, latex dipped articles are susceptible to degradation, both in storage and in service. Some elastomers [e.g., polychloroprene (CR)] are more resistant to degradation, whilst others such as natural rubber (NR) and synthetic *cis*-1,4-polyisoprene rubber (IR) are rather more easily degraded. However, degradation will happen, especially as most dipped latex articles are thin, with a high surface area to bulk ratio, thus exposing most of the bulk of the material to the elements. For this reason, virtually all dipped product formulations will contain an antioxidant. The objective of the manufacturer (and chemical suppliers) is to retard the degradation as much as possible within constraints imposed by manufacturing, cost, safety, storage and end use.

There are many things that can cause the degradation of elastomers. These include heat, light [especially ultraviolet (UV) light], radiation, oxygen, ozone, metal ions (in particular copper, iron, manganese and cobalt) and the effect of chemicals.

There can be several consequences of this degradation, primarily:

- Loss of strength.
- Loss of elasticity.
- Softening and stickiness.
- Hardening and embrittlement.
- Increase in permanent set.
- Cracking.
- Reduction in molecular weight(s) (MW).
- Crosslinking, leading to the formation of a three-dimensional (3D) network.
- Chemical modification of the elastomer chain.

Different elastomers will have different sensitivities to degradation and will respond to it in different ways. Similarly, antioxidants will function differently in different elastomers. For example, Carl [1] reports that some hindered phenolic antioxidants, which are effective antioxidants in NR, can reduce the ozone resistance of CR elastomers. The thermal history of the product, especially during drying and curing, can also affect oxidative stability. Whilst it is frequently necessary to heat the product to ensure the correct level of vulcanisation, excessive exposure to heat during manufacture can reduce long-term stability.

Effective leaching of the product, to remove water-soluble residues, will often help to improve ageing stability.

https://doi.org/10.1515/9783110638097-004

4.2 Mechanism of oxidative degradation

The mechanism of the degradation of elastomers, especially NR, has been extensively studied. A detailed review of the mechanisms involved in the degradation of sulfur- and peroxide-vulcanised NR, and how these studies were carried out, is given by Barnard and Lewis [2]. The polymer network – especially that of sulfur-vulcanised NR – is very complex and can involve many different reactions, with the structure of the rubber network and presence of extra-network materials (fillers, curative residues and so on) all potentially involved, one way or another, in any degradation reactions. For these reasons, only a very simplified picture will be given here, and the interested reader is referred to the extensive literature on the subject for more details.

For thin latex dipped articles, with their high surface area to bulk, the main cause of degradation is likely to be oxidative degradation, which proceeds through autocatalytic attack by free radicals.

A free radical is an atom or molecule with an unpaired valence electron, and is generally formed by the cleavage of chemical bonds. In general, free radicals are highly reactive and unstable, will readily attack double bonds and can initiate fast chain reactions. As mentioned above, these reactions in NR are catalysed by some metal ions such as copper, iron, manganese and cobalt.

However, in a molecule where there is the possibility of electron delocalisation, for example, in a system with conjugated π-electrons, free radicals can be considerably more stable. As mentioned in Chapter 1, the most effective antioxidants tend to be large molecules, with aromatic groups and/or other conjugated structures that can readily delocalise electrons. Thus, an effective antioxidant will give a stable free radical that is much less reactive than those derived from, for example, peroxides.

The reactions responsible for this oxidative degradation are generally considered to progress through three stages: initiation, propagation and termination.

These stages can be simply represented as follows in Sections 4.2.1–4.2.3.

4.2.1 Initiation

Chemical bonds in the rubber hydrocarbon are broken by heat, radiation or chemical activity to form two free radicals (Figure 4.1).

$$R\text{-}R \rightarrow 2R\bullet$$

or

$$R\text{-}H \rightarrow R\bullet + H\bullet$$

Figure 4.1: Examples of initiation reactions forming free radicals (R• is the rubber hydrocarbon and H• is a hydrogen free radical).

4.2.2 Propagation

The free radical can react with oxygen to form a peroxy radical (Figure 4.2).

The peroxy radical can react with rubber hydrocarbon to form an unstable hydroperoxide (Figure 4.3).

$$R\bullet + O_2 \rightarrow ROO\bullet$$

Figure 4.2: Formation of a peroxy radical by the reaction of a free radical with oxygen.

$$ROO\bullet + RH \rightarrow ROOH + R\bullet$$

Figure 4.3: Reaction of a peroxy radical with rubber hydrocarbon, forming a hydroperoxide.

The hydroperoxide molecule will rapidly cleave into alkoxy and hydroxyl radicals (Figure 4.4).

$$ROOH \rightarrow RO\bullet + \bullet OH$$

Figure 4.4: Cleavage of a hydroperoxide into hydroxyl and alkoxy radicals.

These reactions can lead to a rapid increase in the number of free radicals, resulting in a very fast autocatalytic reaction. One of the consequences of these reactions is scission of the polymer chain by these free radicals, and a reduction in MW and physical properties. The free radicals created by chain scission can also react with other polymer chains to create a 3D network that will increase hardness and brittleness with a loss of elastomeric properties. Note that it does not take much in the way of main chain breakage before the physical properties of the elastomer start to be affected.

4.2.3 Termination

These reactions can be terminated in several ways *via* the combination of two free radicals to give more stable products. Examples of termination reactions are shown in Figure 4.5.

In the presence of an antioxidant these active peroxide radicals, responsible for rapid oxidative degradation, can be removed through reaction with the antioxidant (Figures 4.6 and 4.7).

$2\ R\bullet \rightarrow R\text{-}R$

$R\bullet + ROO\bullet \rightarrow ROOR$

$ROO\bullet + \bullet OH \rightarrow ROH + O_2$

$2\ RO\bullet \rightarrow ROOR$

$2\ RHCOO\bullet \rightarrow RC = O + RCHOH + O_2$

Figure 4.5: Examples of different types of termination reactions.

$ROO\bullet + \text{antioxidant} \rightarrow ROOH + A\bullet$

Figure 4.6: Reaction of a peroxy radical with an antioxidant.

and

$ROO\bullet + A\bullet \rightarrow ROOA$

Figure 4.7: Reaction of a peroxy radical with an antioxidant radical.

Free radicals derived from effective antioxidants will be more stable and less reactive than, for example, peroxy and alkoxy radicals. Note also that the antioxidant will be consumed during these processes and sufficient must be present in the formulation to give an adequate reservoir of protection (but not too much, as excess antioxidant can, under certain circumstances, actually accelerate the oxidative degradation of an elastomer). In general, it can be said that chain scission will predominate in elastomers with electron-donating groups (e.g., a methyl group) adjacent to a double bond, as found in NR and IR. Elastomers such as acrylonitrile-butadiene rubber or CR have fewer double bonds, and these are less reactive owing to the presence of electron- withdrawing groups such as chlorine or nitrile adjacent to the double bond. For these elastomers, degradation will result in crosslinking, hardening and embrittlement.

4.3 Mechanism of ozone degradation

Ozone is a highly reactive molecule, which is formed from oxygen by the action of UV light. It is present to some extent in the atmosphere, and can also be formed by electric discharges (e.g., in the vicinity of electric motors) and, in areas where there are high pollution levels, by the action of sunlight on nitrogen dioxide.

All of the elastomers used commercially in latex dipping will have double bonds, to a greater or lesser extent, and therefore can be susceptible to attack by ozone. This susceptibility is greatly enhanced if the elastomer is strained. Ozone will react rapidly with double bonds to form an ozonide (Figure 4.8).

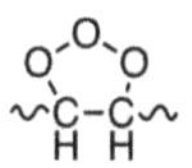

Figure 4.8: An ozonide.

Any such ozonide formed on the surface of an elastomer can easily break, and the elastomer will quickly show the characteristic 'ozone cracking' with a swarm of cracks running perpendicular to the direction of the strain. As with degradation by oxygen, the electron-donating property of the methyl group on the carbon atom next to the double bond in NR and IR makes these polymers particularly vulnerable to ozone attack. Similarly, the electron-withdrawing groups in CR mean that the double bonds in these elastomers are deactivated, and less vulnerable to attack by ozone, giving better ozone resistance to articles made from this polymer.

As ozone will rapidly attack elastomers that are strained, even by small amounts, it is good practice to try and minimise any strained areas in the design and storage of dipped latex articles.

As attack by ozone is a surface effect, antiozonants work by creating a protective layer on the surface of the rubber. Most antiozonants are microcrystalline waxes.

Different formulations and blends of MW are available to fine-tune the performance to the selected application. The waxes for latices are supplied as emulsions that are simply stirred into the latex. The correctly chosen wax will be incompatible with the elastomeric film and will bloom to the surface, creating an ozone-resistant barrier. Obviously this protective layer can be compromised by physical abrasion and flexing, but if sufficient wax remains dispersed in the bulk of the product it can continue to bloom, over time, to the surface and repair any gaps in the wax layer. Typically, antiozonant waxes will be added at concentrations of 1 to 2 phr.

Not only is it important to select the correct antioxidant and antiozonant for the polymer type and service conditions, it is also important to use the correct concentration of antioxidant in the formulation. Not only can antioxidants be expensive materials, but excess antioxidant may react with oxygen to form a hydroperoxide, and hence actually help to promote oxidation of the elastomer. Typically, antioxidants will be used in concentrations of 0.5 to 3 phr, although some applications may require higher levels. It is strongly recommended that the suppliers' recommendations on usage are followed.

References

1. J.C. Carl in *Neoprene Latex*, E.I. Du Pont de Nemours & Co. (Inc.), Wilmington, DE, USA, 1962.
2. *Natural Rubber Science and Technology*, Ed., A.D. Roberts, Oxford University Press, Oxford, UK, 1988, Chapter 13.

5 Reinforcement of latex systems by the use of fillers

5.1 Introduction

Fillers are normally added to rubber latex formulations for two purposes: to lower the cost (usually at the expense of physical properties in the finished product) or to improve the properties. The cost can be lowered by replacing a more expensive material (rubber) with a less expensive material (filler). It is often possible to add a reasonable amount of a cheap filler – say whiting [calcium carbonate ($CaCO_3$)] or lithopone [a mixture of barium sulfate ($BaSO_4$) and zinc sulfide (ZnS)] – in those applications where physical properties are not paramount, and the inherent strength of the base rubber is greater than that required for the end use, for example, in some toys. Improving the properties, typically referred to as reinforcement, is usually a more difficult procedure. Reinforcement is characterised by an increase in physical properties, typically stiffness/modulus, tear strength and tensile strength. The chemical crosslinks in an elastomeric network impart strength and resilience, and it is thought that interaction with filler particles can bring about similar performance improvements. In order for this to happen there must be close physical interactions between the filler particles and the polymer chains, along with a filler particle size that allows this close contact with polymer chains, i.e., the smaller the better [1, 2]. With sufficiently small particles, reinforcement can occur even if the interaction between the filler and the rubber matrix is poor, for example, with graphitised or fluorinated carbon black (CB) [1]. One can hypothesise that the reinforcing effect increases as the filler diameter approaches that of chemical crosslinks. Within the latex dipping industry, many commercially important articles are thin walled and there is thus considerable interest in enhancing the strength of the film, and maybe compensate for any loss of properties brought about by other changes in the product design or formulation. When some properties are enhanced by the use of filler – and this enhancement can be substantial in amorphous elastomers that cannot crystallise on extension – others are frequently adversely affected. For example, the tear strength of a surgeon's glove can be substantially increased by the addition of a suitable filler, but the increase in tensile modulus results in a stiffening effect that can cause unacceptable hand fatigue when the glove is worn for extended periods. Because of this stiffening effect, fillers that give the required reinforcement at low levels of addition are preferred in latex systems. For these reasons, reinforcement is generally applicable to situations where specific targets are required and losses in other aspects of performance can be tolerated. However, some rubber latices – in particular nitrile rubber and synthetic polyisoprene – can give films with a low tear strength, and here the need for reinforcement can be important. So, no matter what the objective, the addition of a filler usually – as

https://doi.org/10.1515/9783110638097-005

with rubber compounding in general – entails some compromises. And when the compounding involves latex an additional problem arises: compatibility of the filler with the latex system.

5.2 Mechanism of reinforcement

A detailed treatment of the mechanism of reinforcement lies outside the scope of this book, and the interested reader is directed to the many excellent treatments of the subject in the literature. However, the basic situation can be described thus. Reinforcement typically is taken to mean improvements in physical properties that result in an increased service life. In the case of tyres this may be increased resistance to abrasion, whereas for industrial gloves it could be increased puncture resistance, for household gloves an increased service life and for condoms a greater resistance to tearing. These are differing requirements that place different demands on the filler, but a very simplified picture of reinforcement can be painted thus. When a strain is applied to a rubber article, interaction between the polymer chains and any filler particles can be broken, and the polymer chains can slip over the surface of the filler. This will help to accommodate and dissipate the stress as the molecular configuration is rearranged and reconfigured when the polymer molecules reassociate with the filler particles. Thus, effective reinforcement requires an efficient transfer of stress to the filler through an intimate filler–polymer interface. This greatly oversimplified picture assumes that the filler is uniformly distributed throughout the polymer matrix and that there is some interaction between the filler and the polymer at the molecular level. In addition to this mechanism, a filler particle at the tip of a crack propagating through the material can force the crack to branch or circumvent the particle. To achieve this requires extra energy to be supplied, and as such imparts reinforcement. The importance of the surface of the filler in interacting with the polymer matrix means that smaller particle size fillers, having relatively larger surface areas, give a greater reinforcing effect. A limit of 500 nm (0.5 μm), above which the filler ceases to give reinforcement, has been proposed [1, 2].

The importance of this intimate contact between the filler and matrix has been demonstrated by many workers [1, 2, 4], and promoting intimate contact through very small filler particle size and improving the compatibility of the filler surface by some type of surface modification has been shown to lead to improved reinforcement [2]. In order to try and promote reinforcement *via* close surface contact between the rubber matrix and fillers in a latex system, there has been interest in developing actual covalent bonding between filler particles and the elastomer matrix [2, 4, 5]. Examples of this approach have included functionalising filler particles by incorporating unsaturation into a polymeric filler or synthesising core-shell copolymers with a rubbery, unsaturated shell on a hard core. This research has shown that the presence of chemical bonds between the matrix and the filler particle appears to have little effect on reinforcement – if anything, the reinforcing effect is reduced when chemical bonding is introduced [2, 4–6]. This supports the postulate that optimum stress relief and reinforcement occur when the rubber chain–filler interaction can be broken when stress is applied, and subsequently reformed [1, 2, 5].

5.3 Comparisons with dry rubber

Dry rubber is famously reinforced by both fine-particle CB and silicone dioxide (SiO_2) (silica). In both cases the increase in physical property – whether in tensile strength, tear strength, puncture strength and so on – resulting from the reinforcement can be substantial. Unfortunately, it is much more difficult with latices. The reinforcing grades of CB and silica have a very small particle size, hence a large surface area. In the dry rubber mixing process, whether by two-roll mill or internal mixer, very considerable shear stresses are imposed on the rubber. These stresses frequently lead to rupture of the polymer chains and the opportunity to force intimate surface contact at the molecular level between the filler and the rubber matrix. In this situation strong interactions can arise between the filler and the rubber. With CB, the interaction can be so strong that rubber is bound to the black in such a way that it cannot be extracted by solvent – the so-called 'bound rubber'. Similarly, the effect of silica fillers in rubber can be enhanced by suitable surface treatment to promote a high level of compatibility between the filler and the rubber matrix [3]. No such shearing and chain rupture takes place during the compounding of latices, and in addition, the latex particle is covered with a protective layer of surface-active materials, adsorbed solvent molecules and other materials such as proteins that stabilise the polymer particles in the aqueous medium. In addition, the rubber phase in a latex elastomer has a high molecular mass, and consequently, a lower molecular mobility. As such, the level of intimate contact described above can only occur during film formation, if at all. It is worth pointing out here that, owing to the mastication procedures in dry rubber, the polymer chains are ruptured and the overall properties of the material are frequently degraded. As a result, an unfilled thin film of natural rubber (NR) prepared from a latex can have a higher tensile strength or tear strength than a film prepared from dry rubber containing a highly reinforcing CB [2].

5.4 The effect of vulcanisation

Much of the work reported on the reinforcement of latices, especially NR latices, has used postvulcanised systems, that is, an unvulcanised, although vulcanisable, latex, which is subsequently vulcanised after filler addition and film-forming. Much of the industry, however, uses prevulcanised rubber latex, where a proportion of the final crosslink density is formed whilst the elastomer is still in liquid form, before the latex is dipped. Prevulcanisation will make the individual rubber particles harder, although the extent will depend upon the formulation and degree of prevulcanisation. Many of the fillers which reinforce postvulcanised latices do not reinforce the same latex when it is prevulcanised. This is thought to be the result of the stronger filler– rubber interactions that are possible with the softer unvulcanised rubber particle [5], whereas the increased hardness of the prevulcanised latex particle can impede the required molecular-level contact. Thus, it is usually easier to reinforce a latex which is subsequently postvulcanised than to reinforce a prevulcanised latex. Nevertheless, reinforcement of prevulcanised NR latices has been found to be possible, especially with fumed silica and some latices of small hard particles such as polyvinyl chloride (PVC), as discussed later in this chapter.

5.5 Materials used as fillers

The literature contains examples of many materials that have been explored at one time or another for use as fillers, for example:
- $CaCO_3$, whiting, chalk.
- Clays of many different types.
- Blanc Fixe ($BaSO_4$).
- Lithopone (a mixture of $BaSO_4$ and ZnS).
- Silica (SiO_2), fumed, precipitated and colloidal, and also other silicates.
- Formaldehyde condensation resins.
- Polymer latices such as high-styrene styrene-butadiene rubber (SBR) latex.
- Dispersions of polymers, such as polystyrene (PS), PVC, polymethyl methacrylate (PMMA), polyurethanes (PU), polyvinyl acetate, polyacenaphthylene and so on.
- Graft polymers.
- Modified starch.
- Proteins.
- Natural polymers such as lignin, chitin, chitosan and shellac.
- Carbon nanotubes and other carbon structures such as buckminsterfullerene.

Most of these are of peripheral importance to the latex industry, and only those materials that have been shown to have industrial potential are considered further in this chapter.

5.5.1 Silicone dioxide (Silicas)

As mentioned above, silica is a very effective reinforcing agent for dry rubber and has been extensively studied in latex systems. However, the large surface area and surface activity of silica particles can cause stability problems by robbing the latex particles of stabiliser entities, and incorporating silica into latex systems will usually require some attention to the stabiliser system. Both precipitated and fumed silica have been used as fillers in latex. Precipitated silica is formed by the reaction of an alkaline silicate solution with a mineral acid, which gives a porous particle of SiO_2 with a size range of 5 to 100 nm, although these particles tend to exist as agglomerates of 1 to 40 µm in size. Fumed silica is prepared by the high-temperature pyrolysis of a silicone-containing material, such as silicon tetrachloride. Particles of fumed silica are non-porous with a size range of approximately 5 to 50 nm. The fumed silica has found more widespread use in latices. Some of these silicas are available as aqueous dispersions [7], simplifying the compounding stage, providing that the usual precautions of matching pH and careful addition are followed. Fine-particle silicas can increase the tensile strength and tear strength of films from latex, even at fairly low concentrations (2–5 phr), although these increases are not always maintained after accelerated ageing or chlorination. However, an increase in modulus/stiffness will almost invariably result from adding silica, and such increases in stiffness are generally unwelcome, especially in articles where flexibility is important, such as gloves and condoms. As with other fillers there is a difference in the reinforcing ability of prevulcanised and postvulcanised latices, although unlike some other materials, silica can give good levels of reinforcement in a prevulcanised latex. The structure of the latex film and the extent of agglomeration and 'clumping' of the silica after addition can depend upon the way in which the silica has been added, and the method of film formation, for example, straight dipping or coagulant dipping. Agglomeration can also be influenced by whether the silica is added before or after any prevulcanisation. It has been found that, generally, the reinforcing effect is greater with straight dipped films and with the silica added at the final compounding stage. The presence of silica as a filler in NR articles is also claimed to reduce the amount of extractable latex proteins by adsorbing them onto the active surface of the silica [7], and other studies have shown that retention of physical properties during accelerated ageing can be higher in silica-filled elastomers [7, 8].

As well as adding the silica as an aqueous dispersion, it can be formed as fine particles *in situ* by various reactions. These include the so-called sol-gel reaction, where a silica precursor, such as tetraethoxysilane, is incorporated into the rubber. Tetraethoxysilane is non-polar and easily incorporated into the rubber, even into a finished article, by swelling. Hydrolysis of the tetraethoxysilane will then give finely dispersed silica particles within the rubber matrix [9]. Similarly, SiO_2 can be formed *in situ* by incorporating a solution of sodium silicate (Na_2SiO_3) into an anionic latex. The Na_2SiO_3 solution is alkaline and easily incorporated into the latex; however, if the pH is reduced – which can happen as the latex film is dried – the Na_2SiO_3 will convert to silica [10].

5.5.2 Clays

Another filler type which has stimulated great interest is fine-particle clays, which include clays of the smectite class, such as sodium montmorillonite (Na-MMT) or fluorohectorite. They exist in the form of layered silicate nanoparticles approximately 1 nanometre (10^{-9} metre) thick and up to approximately 2,000 nanometres (2 µm) in length and width. These large aspect ratio plates form laminar layers, which can be swollen by hydration and incorporated into rubber latices by simple mixing. They usually contain sodium or potassium ions that can aid their integration into an anionic latex. The montmorillonite (MMT) materials can be modified by cation exchange with organic ammonium salts to give anisotropic layered plates with improved compatibility and dispersibility in latex systems [11].

When clays are dispersed properly in a latex system, two types of incorporation are possible: intercalation, where the polymer diffuses between the silicate layers, but the silicate layers retain some short-range order, and exfoliation, where the silicate layers are fully separated and dispersed in the latex. It is generally true that rubber blends where the clay is intercalated have lower permeability to gas and liquids, whilst exfoliation of the clay gives higher modulus and strength. When preparing these systems, stability of the latex can be a problem, with thickening and gelation of the latex reported. Correct preparation of the clay dispersion and the correct choice and formulation of latex stabiliser system are essential prerequisites. When correctly prepared, the clays can be shown by electron microscopy and X-ray dispersion studies to be well dispersed in the latex. NR latex articles incorporating these layered silicate nanoparticles have been reported with varying effects on the physical properties of the composite. Most workers agree that the permeability to gases and liquids is reduced, but there is less agreement amongst the reported results on physical properties. Varghese and co-workers [12] have compared sodium fluorohectorite and sodium bentonite with a commercial (non-layered) clay in NR, to form vulcanised NR–clay nanocomposites. The clays were compounded into the latex at a level of 10 phr by simple mixing. The highest tensile strength was obtained on the film containing fluorohectorite, although the differences between the different composite films were small. In a similar fashion, the results for elongation at break were comparable for all of the composites, except for the composite containing fluorohectorite, which was lower. The clay fillers all increased the modulus of the films, with fluorohectorite having the greatest effect. The study included the dynamic properties of the composites, and those films containing the layered silicates had a greatly increased storage modulus and a greatly reduced loss factor (tan δ). This was interpreted as showing that there was strong adhesion between the rubber and the silicate layers, especially in the case of sodium fluorohectorite. Other types of rubber latices have been compounded with layered silicates of this type, but the general conclusions are similar to those found for NR.

There has been considerable work looking at the use of layered silicate clays. Work on a commercial prevulcanised NR latex containing 5 phr of MMT or sodium fluorohectorite [13] showed that the tensile strength was only marginally increased and the ageing resistance (admittedly at 100 °C) was poor. The tear strength and modulus were also increased, although only modestly. Both of these clay materials were found to adversely affect the latex stability, leading to significant increases in viscosity and coagulation. Na-MMT was found to result in a considerable increase in tensile strength when used at lower concentrations (3 phr) in NR latex before vulcanisation [14]. A tensile strength of 31.5 MPa was reported, compared with 20 MPa on the control (unfilled) sample. The values after ageing at 70 °C were even more impressive, at 39.4 and 19.0 MPa for sample and control, respectively. The elongation at break of the samples was also very high. As seems almost inevitable, the addition of the filler also increased the modulus.

A different procedure for incorporating reinforcement was carried out by Liu and co-workers [15], who prepared nanocomposites of NR with organoclay by intercalating butyl acrylate (BA) into the clay layers and then grafting the BA onto the NR backbone. This grafting procedure is claimed to give higher physical properties and glass transition temperature than for the NR and ungrafted composite controls.

A practical application of the use of clays as a filler has been reported by Goldberg and co-workers [16], who prepared an aqueous dispersion of vermiculite and blended it with butyl rubber latex. The composite latex has an exceedingly low permeability and has been used in the manufacture of tennis balls.

5.5.3 Other inorganic materials

Most inorganic materials other than the layered silicate clays or silicas described above are non-reinforcing fillers, but some workers have reported significant levels of reinforcement using fine $CaCO_3$. Manroshan and Baharin [17] examined the effect of nanosized acrylic-dispersed $CaCO_3$ in prevulcanised NR latex films. As is usual, the modulus increased as the loading of filler increased, and tensile strength and elongation attained maximum values at 10 phr of $CaCO_3$. The physical properties improved with accelerated ageing, suggesting that the heat treatment gave a more intimate contact between the filler and the matrix. Other studies with a nanosized $CaCO_3$ [13] at 5 phr in NR latex gave slight increases in tensile strength to 22.7 MPa (initial) and 23.5 MPa (aged).

Claramma and co-workers [8] evaluated the effect of whiting and china clay in a prevulcanised NR latex, as well as the effect of precipitated silica. The silica and china clay improved the tear strength of the rubber, whereas the whiting had no effect. All of the fillers increased the modulus and reduced the tensile strength and elongation, although up to 10 phr of the silica improved the resistance of the rubber to accelerated ageing.

5.5.4 Polymeric particles

Polymeric materials such as latices of PS, PMMA, PVC or high-styrene SBR have been found to reinforce natural and styrene-butadiene latices [8]. Hard polymers reinforce more than softer polymers, although the polarity of the polymer does not seem to be important [2]. Significant increases in tear strength can be obtained in this way, although the magnitude of the increase is again greater with a post-vulcanised system. As with many other types of reinforcing fillers, chlorination and accelerated ageing can reduce or wipe out any gains, and as usual, the modulus of the rubber is increased by this blending. As mentioned earlier, the mechanism for reinforcement of dry rubber presupposes an intimate contact, even covalent bonding, between the filler and the rubber matrix, and one might think that introducing chemical bonds between the elastomer matrix and the filler would be of benefit in a latex system. These bonds can be facilitated in a number of ways, such as incorporating an element of unsaturation into a PS latex by copolymerising with a small amount of divinyl benzene or using a covulcanisable elastomer such as a high-styrene SBR as a filler. Alternatively, a core-shell polymer latex can be synthesised, using a hard polymer such as PS or PMMA as the core and a vulcanisable elastomer such as polyallyl methacrylate or polyisoprene as the shell [2]. However, investigations have shown that there is no improvement from functionalising the latices in this way – in fact if anything, the tear strength and tensile strength were reduced slightly, although promoting more intimate contact between the filler and the elastomers with a suitable shell polymer, such as polybutyl acrylate, did give a slight increase in tear strength over that obtained with the filler alone [2]. Thus, it appears that compatibility to the point of chemical bonding will not necessarily give the optimal reinforcement. As discussed above, it is thought that this is best achieved by a system where the stresses can be relieved by filler/matrix slippage, followed by the reestablishment of intimate physical interactions [1, 3, 5].

5.5.5 Graft copolymers

The ultimate in compatibility between filler and elastomers might be thought of as actually incorporating the filler into the polymer backbone. Reinforcement is generally thought to be enhanced by small hard domains interfacing intimately with the elastomer network, which are able to dissipate the energy when the matrix is strained. Such small hard domains can be prepared on a molecular level by the grafting of a glassy polymer – such as methyl methacrylate (MMA), butyl methacrylate or styrene – onto the polymer backbone [18]. NR latex with MMA grafted onto it is commercially available as MMA-grafted latex, and blends of MMA-grafted latex with NR latex give latices with enhanced tensile, tear and puncture strengths [18].

Grafting of other monomers onto NR latex has also been achieved [18], although MMA-grafted latex is the only one to have achieved significant commercialisation.

5.5.6 Polyurethane dispersions

As has been consistently reported, the addition of a reinforcing filler will increase the modulus of the system. Although this is generally regarded as undesirable in many dipped latex articles, one area where this modulus increase is sought after is in the development of thin condoms. As thinness is associated with sensitivity by condom users, a goal frequently sought by condom manufacturers is to produce the thinnest condom possible. However, condoms are classed as medical devices, and have to meet a minimum bursting pressure to comply with the requirements of the international condom standards. One of the consequences of making the condom thinner is a reduction in burst pressure. This effectively imposes a minimum thickness on condoms if compliance to the standard is to be achieved. Choosing a suitable reinforcing filler for the condom will increase the modulus, and therefore should increase the burst pressure and allow a thinner condom to be made, with the added benefit that tear strength is also often improved. Unfortunately, another mechanism is also involved. Adding a reinforcing filler can also reduce elongation at break and in a condom burst test this will be seen as a reduction in burst volume. Condom burst volume and burst pressure are linked: the greater the volume, the higher the pressure and *vice versa*. Thus, the consequence of increasing condom modulus is usually an increase in inflation pressure at any point along the pressure/volume curve, but the lower burst volume means that the ultimate burst pressure is usually no higher than an unreinforced condom, and may be lower (Figure 5.1).

However, one type of filler system has been patented [19] which defies this behaviour. Certain types of PU dispersion can be added to NR latex, and will give increased condom burst pressure (and some increase in tear strength) without the same level of reduction in burst volume. Thus, a route is opened to make thinner condoms that comply with the burst requirements of the standards.

5.5.7 Proteins

NR latex contains proteins, and the film-forming process results in these proteins being forced together into domains that can then act as particulate fillers [20]. The reinforcing effect of these proteins is one reason why NR has a considerably higher tear strength than synthetic polyisoprene or deproteinised NR. It is also possible to add proteins as reinforcing fillers, for example, Jong [21] found that soy protein, in a variety of forms, can reinforce products

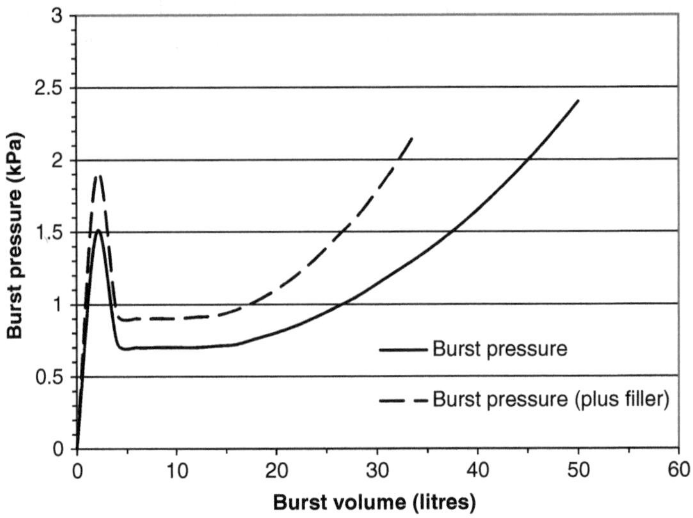

Figure 5.1: A representation of the burst pressure/burst volume relationship for a reinforced and an unreinforced condom.

dipped from SBR latex. Starch, in various modified forms has also been found to be of value as a reinforcing filler [22, 23]. The prospect of removing the potentially harmful latex proteins from NR, and replacing them with a more benign protein that will reinforce in a similar way, is an attractive one that merits further study.

5.6 Conclusions

Reinforcement of latex dipped articles is a very desirable goal, both for improvement in physical properties and possible reduction in costs; however, this goal remains elusive. Recent work in the use of layered silicates has shown that these materials can greatly reduce permeability and offer some improvement in physical properties, such as tear strength, to non-crystallising rubbers such as SBR latex. As with dry rubber, the effect of particle size is important – the smaller the better, even going down to molecular levels, and an intimate contact between filler and matrix is essential. Bonding between the matrix and the filler is not required, however, and can even be detrimental. The effect of blending PU dispersions into latex is interesting and gives results that are outside of the normal experience. Perhaps there is more to be uncovered in this system. The use of graft latices, such as MMA–NR latices, appears to be underexploited, bearing in mind what they can offer. However, in almost all cases, it does not appear to be possible to gain in one area without losing in another. In the vast majority of cases, any reinforcement will increase modulus and decrease elongation. Useful enhancements can be achieved, but the material that lifts all of the critical properties onto a higher plane has yet to be found.

References

1. G.R. Hamed, *Rubber Chemistry and Technology*, 2000, **73**, 3, 524.
2. K.A. Popoola in *Mechanical Reinforcement of Films from Rubber Latices by Added Polymer Particles*, London School of Polymer Technology, The Polytechnic of North London, UK, 1988 [PhD Thesis].
3. E.P. Pluedemann in *Silane Coupling Agents*, Plenum Press, New York, NY, USA, 1982.
4. M. Morton, J.C. Healy and R.L. Debecour in *Proceedings of the International Rubber Conference*, Brighton, UK, 1967, p.175.
5. D.C. Blackley in *Polymer Latices: Science and Technology*, Volume 3, Chapman & Hall, London, UK, 1997.
6. D.C. Edwards, *Journal of Material Science*, 1990, **25**, 10, 4175.
7. M.D. Morris and J. Anand in *Proceedings of the 150th ACS Rubber Division Meeting*, Louisville, KY, USA, 1996.
8. N.M. Claramma, L. Varghese and N.M. Mathew, *Journal of Elastomers and Plastics*, 1999, **31**, 130.
9. S. Kohjiya and Y. Ikeda, *Rubber Chemistry and Technology*, 2000, **73**, 3, 534.
10. D.M. Hill, inventor; LRC Products Ltd., assignee; EP1778772B1, 2016.
11. M. Ganter, W. Gronskl, P. Reichert and R. Mülhaupt, *Rubber Chemistry and Technology*, 2001, **74**, 2, 221.
12. S. Varghese and J. Karger-Kocsis, *Polymer*, 2003, **44**, 17, 4921.
13. S. Awang, K. Vivaygananthan, S. Bahri, A.R. Amir Hashim and M.Y. Amir Hashim in *Proceedings of Latex 2006*, Frankfurt, Germany, 2006, Paper No.19.
14. A.S. Aprem, A. Jacob and S. Nath Pal in *Proceedings of Latex 2006*, Frankfurt, Germany, 2006, Paper No.22.
15. L. Liu, Y. Luo, D. Jua, W. Fu and B. Guo, *Journal of Elastomers and Plastics*, 2006, **38**, 2, 147.
16. H.A. Goldberg, C.A. Feeney, D.P. Karim and M. Farrell in *Proceedings of Latex 2004*, Hamburg, Germany, 2004, Paper No.19.
17. S. Manroshan and A. Baharin, *Journal of Applied Polymer Science*, 2004, **96**, 1550.
18. D.S. Campbell in *Natural Rubber Science and Technology*, Ed., A.D. Roberts, Oxford University Press, Oxford, UK, 1988, Chapter 14.
19. D.M. Hill and C. Brodin, inventors; LRC Products Ltd., assignee; US9173767B2, 2015.
20. K.L. Chong and M. Porter in *Proceedings of the International Polymer Latex Conference*, The Plastics and Rubber Institute, London, UK, 1978.
21. L. Jong, *Applied Science and Manufacturing*, 2006, **37A**, 3, 438.
22. C. Llu, Y. Shao and D. Jia, *Polymer*, 2008, **49**, 8, 2176.
23. S.C. Petersen and L. Jong in *Proceedings of the ACS Polymeric Materials: Science and Engineering – Fall Meeting*, Boston, MA, USA, 2007.

6 Latex stability and film-forming

6.1 Latex stability

6.1.1 Introduction

One of the key elements in latex dipping is the stability of the latex system. If the system is not stable enough then the various external influences on the latex – shear forces in pumping or agitation, elevated temperatures and so on, will cause coagulum to form in the latex. If the latex is too stable then there will be too much flow of the latex when it is picked up on the former – both with straight dipping and coagulant dipping – and the subsequent latex film will be uneven, with the risk of cracking. It is also messy. So the ideal stability for dipping is balanced precariously on the boundary between stability and instability.

What is meant by stability in this context? A latex system contains a myriad of small particles, and the total free energy of the system would be greatly reduced if these particles were coalesced together into one mass. Thus, a latex system is intrinsically in a high-energy state, and nature, as always, will be trying to reduce the free energy of the system. To try and put this into context, for example, in one teaspoonful (approximately 5 ml) of natural rubber (NR) latex at 60% dry rubber solids there will be around 6×10^{12} (six million million) individual particles. Energised by the ambient temperature, these particles are in continual motion and if they collide with each other they may coalesce and form larger particles. The system is maintained in its dispersed state by the potential energy barriers that exist between the particles. Stability is conferred on the latex by manipulating these potential energy barriers so they prevent the rubber particles coming so close to each other that they undergo such collisions.

This can be illustrated in a simplified form by the familiar free-energy curves shown in Figure 6.1.

The X-axis represents the separation between particles and the Y-axis represents the free energy. Curve (1) shows the effect of the combined repulsive forces in the system. As the particles approach each other, the free energy of interaction of the surfaces increases, i.e., these repulsive forces mean that more energy has to be applied to force the particles closer together. However, in addition to the repulsive forces in the system, we also have attractive forces working to bring the latex particles together. There are several different types of attractive forces, collectively known as the van der Waals forces. The effect of these is shown as curve (2). As the particles approach each other the free energy of the system decreases – an energetically favoured situation. It is generally assumed that the net forces are additive, and curve (3) shows the sum of the attractive and repulsive forces. For stable latex systems, there is a maximum in the free energy, which appears as a 'hump' or

https://doi.org/10.1515/9783110638097-006

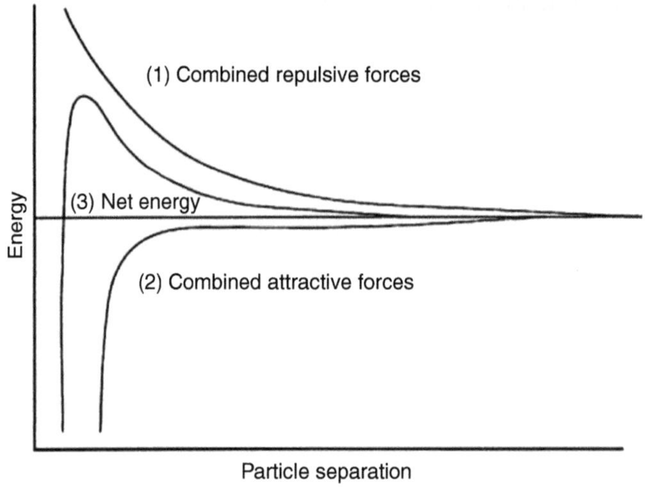

Figure 6.1: A representation of the forces working on a latex system.

energy barrier at a particular interparticle distance. If the particles are able to approach closer than this distance, coalescence will rapidly occur and the free energy of the system will be reduced. This is what nature is trying to achieve. The size of this energy barrier is an indication of the stability of the system, or alternatively, the amount of energy needed to bring the latex particles close enough together to induce coalescence.

The problem in compounding a latex is, quite simply, to ensure that this energy barrier is appropriate for the designed purpose: high enough so that the latex can be handled during the compounding operation – pumping, transfer, compounding, heating, agitation and so on. The latex needs to be stable enough in the dipper to resist the agitation and thermal stresses imparted by hot formers and cooling coils, to resist any stray drops of coagulant, to be slow to cream and to counter any instability caused by some of the compounding ingredients. Yet at the same time, it has to give a coherent and firm gel when coagulated, or set up rapidly and not give excessive flow when straight dipped. In addition, any stabilisers must not interfere with the film-forming process.

Of the two effects operating on the latex particles (attraction and repulsion) it is much easier to modify those operating on repulsion. This will be achieved through the stabiliser, or stabiliser combination in the latex formulation. Note also that diluting the latex will reduce the likelihood of interparticle collision, and thus water can be regarded as having a stabilising effect.

The science behind latex stability is complex and only a very much simplified picture will be presented here. Much more rigorous and indepth treatments are given in the literature, for example, volume one of Blackley's book on polymer latices [1] and the excellent introduction to colloid science by Everett [2].

Current thinking proposes four different types of repulsive forces acting on latex particles, which can influence and control latex stability. The overall stability is likely to be the result of a mixture of some or all of them.

6.1.2 Electrostatic repulsion

The first, and arguably most important, repulsive force is electrostatic repulsion. This force can only act in an aqueous medium and is the result of the electrical charge density on the surface of the latex particles. The latices used for dipping carry an electrical charge – almost always a negative charge – on the surface of the latex particles, brought about by the adsorbed ions or the ionisation of chemical groups on the particle surface. The electrical charge on these ions gives rise to a cloud of ions of opposite charge (counterions) around the particle to try and preserve electrical neutrality. As two particles approach each other, their respective clouds of counterions will try to interpenetrate, which will be resisted as the counterions carry the same charge and will thus repel each other. This electrostatic repulsion is highly dependent on the pH and ionic strength of the dispersion medium – water in the case of latices. Whilst increasing the pH of an anionic latex will reinforce stability by enhancing the electrostatic charge, note that an increase in the ionic strength of the medium, for example, by adding salts, will compress this electrostatic layer and decrease the charge density. Materials which contribute greatly to electrostatic stabilisation are those that can exist as charged entities, for example, anionic surfactants and materials such as proteins. The hydrophobic part of the surfactant (e.g., the hydrocarbon chain in a fatty acid soap) can associate with the particle surface whilst the hydrophilic end of the molecule gives compatibility with the aqueous medium. Alkali will help to ensure that these materials are maintained in an ionised state so that this electrostatic repulsion is effective.

6.1.3 Steric repulsion

A second repulsive force that can affect and control latex stability is steric repulsion. This is the result of macromolecular chains adsorbed or attached to the latex particle, and which extend into the dispersion medium. As the latex particles approach each other, these chains provide a barrier to prevent the particles approaching closely enough to coalesce. Any entangling of these chains will reduce the entropy of the system and increase the free energy, which will be resisted. In addition, an osmotic effect can drive the dispersion medium into the spaces between the chains, helping to keep them apart. To maximise this effect, the macromolecule needs to have a segment or segments that can associate closely with the surface of the latex particle, and other segments that can associate with the dispersion medium. It is obviously important that these macromolecular chains do not desorb or collapse under the conditions existing in the latex. This repulsive effect is important for both aqueous and non-aqueous systems. Stabilisation by steric repulsion is relatively little affected by ionic strength and pH, (although if changes in the ionic concentration lead to the chains adopting a tighter conformation, the magnitude of the stability enhancement will diminish), but can be highly temperature dependent. This temperature dependency is not consistent e.g., the stability conferred by these effects can be increased initially by a temperature rise, but there may then be a reduction in solvency of the macromolecules and other entropic effects. Steric stability is conferred on a latex by materials such as surfactants (both anionic and non-ionic) and other macromolecules such as proteins and water-soluble polymers. Note that these stabiliser molecules usually have to displace other entities to become adsorbed onto the particle surface, and so can take some time to achieve the maximum stabilisation effect. This is thought to be one of the mechanisms that occur in any maturation period after compounding, and helps explain why maturation will improve the properties of a compounded latex.

6.1.4 Solvation stabilisation

A third repulsive force which may be important at times in conferring stability is that brought about by solvation of the latex particle surface. In this situation, molecules of the dispersion medium – which in the case of latex dipping is almost invariably water – can be closely associated with the particle surface. Any such 'shell' of water molecules can provide a barrier to particle coalescence. In addition, the surface of this solvated particle will be similar to the dispersion medium, and so the free energy at the interface will be decreased, thus reducing any thermodynamic advantages in particle coalescence. Solvation of the latex particle surface is helped by surface characteristics that are compatible with the dispersion medium – so again, materials such as surfactants and proteins are important in maintaining and controlling stability.

6.1.5 Depletion stabilisation

A fourth type of repulsive force is that resulting from depletion or exclusion effects, which arise from the presence of macromolecules dissolved in the dispersion medium. In a dilute solution, a dissolved macromolecule can adopt any conformation. However, in a latex system, if particles of the dispersed colloid approach such a macromolecule, they can start to exclude the macromolecule from certain conformations. Such a restriction will be resisted and will contribute to the repulsive force between the particles. However, if the particles are able to approach sufficiently to exclude all macromolecules from the space between them, so that there is only pure dispersion medium between them, then further mutual approach will expel dispersion medium, dilute the adjacent medium and reduce the concentration of macromolecules, a thermodynamically-favoured process. In this case, the coalescence will then be energetically favoured. Such dissolved macromolecules in a latex system will again include materials such as surfactants and proteins.

6.1.6 Controlling latex stability

There are a selection of ways in which the stability of the latex system can be engineered, depending upon the end use of the latex and which stabilisers are used and how. Concentrated NR latex already has a level of stability, brought about by naturally occurring proteins and higher fatty acid soaps derived from alkaline hydrolysis of the phospholipids that are present. This inherent stability may be sufficient for some applications, without the need for any other added stabilisers. However, in many cases latex stability will need to be enhanced to ensure a good and consistent level of quality. Typically, the level of alkalinity will be controlled to maintain and reinforce the electrical double layer around the particle, but note that with a carboxylated latex, such as many nitrile latices, additional alkali can promote the ionisation of the carboxylic groups, which in turn can lead to an unacceptable increase in viscosity. In polychloroprene (CR) latex, the alkali also performs an important function as an alkali reserve to counter the effect of any release of hydrochloric acid through hydrolysis of the base polymer, which would destabilise the latex. Because of this high level of alkali, CR latices can be very stable and may actually need a controlled level of destabilisation before they are suitable for dipping. Added surfactants are widely used to control stability, and many different types are available and used. Because of the (generally) anionic nature of latices used for dipping, anionic surfactants such as long-chain carboxylates, sulfonates and sulfates can be appropriate. These materials are also foam promoters, as well as latex stabilisers, and care should be taken not to introduce air into the latex during mixing. Fatty acid soaps are widely used, but the number of carbon atoms in the chain can be of importance, as discussed in Chapter 1.

Latex compounds can also be stabilised by water-soluble hydrocolloids, such as casein, alginates or cellulose derivatives. By using these different types of stabilisers, the different mechanisms responsible for latex stability can be brought into play, or suppressed, to give, one hopes, the correct balance of stability for the latex and process in use.

Finally, some of the synthetic latices are carboxylated and these carboxylic acid groups can play a major part in stabilising the polymer chain. This reduces the need for added surfactants to stabilise the latex, but can result in the finished article having some degree of water sensitivity.

6.1.7 Measuring latex stability

Quantifying the effect of stabilisers on latex formulation development in the laboratory is difficult. Unfortunately, stability issues (both insufficient stability and excessive stability) are sometimes not apparent until the latex is on the manufacturing plant. Measuring latex stability can be carried out in two ways: measuring the amount of energy needed to surmount the potential energy barrier to create particle coalescence – the mechanical stability, and reducing the barrier to coalescence by chemical means – the chemical stability. Standard methods exist [3–5] to measure the mechanical stability of natural and synthetic latices, and although they are not without problems, they are generally widely accepted and used. Chemical stability is much more difficult to assess. Several tests exist, which in the main rely upon reducing the latex stability by the addition of chemical destabilisation agents, for example, zinc oxide (ZnO) or zinc acetate, and then measuring the mechanical stability or changes in viscosity. However, the way in which one latex behaves towards a test may well be completely different to the way in which another latex behaves. A comparison of several of these chemical stability tests has been made by Blackley and Ong [6], who found that most of the tests were reasonably concordant in their ranking for the NR latices examined. However, it was considered that the tests using mechanical stability to assess the effect of the added destabilising agent were doing little more than providing an indirect measurement of mechanical stability. Pendle and Gorton [7] evaluated the chemical stability, or 'processability' of some 40 NR latex concentrates, and concluded that the tests based on changes in mechanical stability and those based on viscosity changes measured different aspects of latex stability. Unfortunately, the only reliable way to assess the chemical stability of the latex is empirically – put the formulation on the plant and dip it.

6.1.8 Using latex stability

The correct balance between the stability and instability of the latex will have a major impact on the ease with which the latex forms a coherent, strong, flaw-free film, and it therefore has a major impact on the quality of the products being dipped. Straight dipping requires that the latex viscosity increases substantially as soon as possible after withdrawal from the dipper, so that the latex will not flow excessively, and give uneven pick-up and cracking. In this case, it could be appropriate to have a substantial contribution from electrostatic stabilisation, using a fugitive alkali such as ammonia that can be rapidly lost from the surface of the film after dipping, as well as any surfactant.

Stability of a latex designed for coagulant dipping would be more likely to use a fatty acid soap of the appropriate chain length, to give sensitivity to the coagulant and a firm gel. In both cases, there may well be a small amount of a non-ionic surfactant, such as ethylene oxide condensate, to give a measure of background stability that will not be affected by ZnO or coagulant.

6.1.9 Zinc oxide thickening

In a formation containing zinc ions – and virtually all latex formulations will contain zinc compounds – be aware that too much ammonia may lead to instability through the reaction of zinc ammines. This can lead to the phenomenon known as 'zinc oxide thickening', especially in an ammonia-preserved NR latex. A general formula for zinc ammines is shown in Figure 6.2. It is thought that in NR latex containing ammonia the tetraammine (where the value of n is 4) predominates [1].

$$Zn\,(NH_3)_n^{2+}$$

Figure 6.2: General formula of a zinc ammine ('n' can take any value from 1 to 4).

The zinc ammine can decompose, releasing zinc ions into the aqueous phase. These zinc ions can then react with fatty acid soaps to give an insoluble zinc salt. Thus, any stability conferred by the fatty acid soaps will be removed and the latex will thicken or even coagulate. Decomposition of the zinc ammine is accelerated by heat and this ZnO thickening reaction can be of value in heat-sensitised dipping. Zinc-mediated thickening can be reduced by the use of potassium hydroxide, and so the balance between fugitive and fixed alkali will be important.

6.1.10 Final thoughts on latex stability

Whilst it is hoped that the information given in this section will help latex formulators in their tasks, three things should be noted: small changes in some stabilisers – especially some of the non-ionic surfactants – can have large effects. Some stability issues can also take time to show themselves: the author has known a reformulated latex stability system that worked well initially, but after some 7–10 days on the dipping line gave rise to an unacceptable level of coagulum. Finally, stability issues are difficult and complex. If your formulation works well, don't be tempted to change it without good reason.

6.2 Latex film-forming

6.2.1 Introduction

Latex film-forming is fundamental to the dipping industry. If the latex does not form a strong, coherent film then useable products cannot be made. The theory of film-forming and the forces acting on the latex particles are complex, with several different theories through the ages being proposed, discussed and at times rejected. Even after decades of study, there is still no agreement over the mechanisms involved. The interested reader is referred to Blackley's book [1], which contains details of the various theories and mechanisms, and there is a comprehensive overview by Steward and co-workers [6]. This section will not attempt to go into these theoretical considerations, but will give a brief and greatly simplified view of the process for those working with latices on dipping lines, in process control laboratories and development laboratories.

A prerequisite for successful latex film-forming is that the substrate is properly wetted by the latex and any coagulant, and it is assumed here that this is the case. Wetting and its application to latex dipping is discussed in Chapter 8.

6.2.2 Film-forming

Latex film-forming takes place when the small particles of the polymer come together and fuse into a coherent mass. This requires that the forces separating the particles, and giving the latex its stability, are overcome. This is usually accomplished by evaporation of the water or by chemical destabilisation of the system. When this happens the particles are forced together, and if certain conditions are met, a film will be formed. Amongst these conditions is the minimum film-forming temperature (MFFT), that is, the minimum temperature at which the polymer particles will coalesce together. It is similar to, but not necessarily the same as, the glass transition temperature (T_g) and can be altered to some extent by the presence of plasticisers and, possibly, by particle size. Values of MFFT are not readily accessible in the literature, so Table 6.1 lists the T_g of common polymers used in latex dipping as an aid to understanding. Homopolymers of some of the monomers used in the synthesis of synthetic latices are included for comparison. It can be seen that neither polyacrylonitrile (PAN) nor polystyrene (PS) will form films at room temperature, but their T_g, and hence their film-forming ability, can be modified by copolymerisation.

With straight dipping, as the water evaporates from the wet film, the volume is reduced and the particles are forced closer and closer together, increasing the frequency of collisions. This is initially resisted by the stabiliser system, which is, of course, as described earlier, designed to keep the particles apart. However, eventually the forces acting on the particles as they are gradually pushed together will overcome the stabilising forces, and the particles will start to aggregate. This will typically happen as the volume fraction of the polymer approaches 60–70%. As the drying proceeds this aggregation of the particles will become an irreversible coalescence of the individual particles and the interparticle boundaries will start to disappear. The aggregation

Table 6.1: T_g of some polymers.

Polymer	Approximate T_g (°C)
NR*	−70
Acrylonitrile-butadiene rubber	−60 to −10, depending on the formulation
PAN	90
Polybutadiene	−85
Styrene-butadiene rubber	−70 to 50 depending on the formulation
PS	95
CR	−45

*Synthetic *cis*-1,4-polyisoprene will have a very similar T_g

process will now be energetically favoured, as the increase in particle size with coalescence will reduce the surface area of the particles, and result in a loss of the surface free energy of the system. As coalescence progresses, any substances incompatible with the bulk polymer will be expelled, together with residual water, towards the film surface. Typically, these residual materials will include the stabilisers, such as soaps and proteins. With coagulant dipping the stabilising materials will have been partially or completely removed through chemical reaction with the coagulant, and the barriers to particle integration reduced, leading to rapid coalescence of the particles.

There is now a phase inversion and a continuous film of the polymer starts to form. In an aqueous latex system, the water is the continuous phase, with the polymer particles as the discrete phase. As coalescence proceeds, the polymer becomes the continuous phase with the water located in domains within the polymer. At this point (known as the 'wet-gel' point) the film will have a measure of strength, although well below that which will eventually be attained. The wet-gel strength is highly dependent on the base polymer as well as the dipping conditions. NR typically can have a good wet-gel strength, whilst that of nitrile rubber can be very weak.

At this stage there are still clear interparticle boundaries (under appropriate magnification), although these boundaries will gradually disappear to a greater or lesser extent as the film matures. Of course, the non-rubber materials – the surfactants, colloid stabilisers, products from reaction with any coagulant and other components of the system – are still present. Some of these will be forced out as the rubber particles fuse together, whilst others will remain trapped within the matrix or dissolve within the polymer. These included materials can hinder good particle-to-particle contact, and thus weaken the resulting film. In most cases it is better to remove them, and this is where leaching of the film can play an important role. A thorough leach, dissolving out the water-soluble residues, will usually improve the strength of the film, although of course, the leach water will need to be changed at intervals to prevent leach efficiency being reduced by a build-up of extracted chemicals.

This process of particle coalescence and integration is affected by a whole range of outside influences, including temperature, particle size and particle size distribution, and the amount and type of stabiliser present. In general, any factor that can increase the mobility of the polymer chains will increase the ease of film integration, and conversely, anything that reduces the chain mobility (such as crosslinking or chain branching) will impede film-forming. Thus, the extent of any prevulcanisation will play a part, as will the molecular mass of the polymer – a lower molecular mass will give faster particle integration. As stated above, plasticisers will affect the film-forming, but note that the plasticiser might not be deliberately added. Under certain conditions water can plasticise the polymer, especially if there are other materials present that can promote association between the water and the polymer. Reducing particle size also facilitates interparticle coalescence. The fact that prevulcanised latices can give good coherent films is of interest, as usually the presence of crosslinks within a polymer restricts viscous flow of the material, which is a requirement for good interparticle coalescence. However, as is well known, NR latex in particular can be prevulcanised to give quite a high degree of crosslinking within the particle, and yet these particles will still coalesce to give a strong and resilient film.

The rate of drying of a latex film is initially rapid, but slows down as film integration occurs, and the water molecules increasingly have to diffuse between the particles of rubber as they integrate. The characteristics of the polymer become increasingly important, and articles dipped from materials with a low permeability [such as isoprene-isobutylene (butyl) rubber] are thus very difficult to dry unless the film is very thin.

This particle integration is not always complete on a finished dipped latex film. Interfaces between latex particles may still be seen on dried films when examined under high magnification, although the material can still have a high strength. These boundaries can gradually disappear as the film matures over the subsequent days and weeks. As this process continues, the rubber particles can diffuse into one another and gradually lose their identity, and we can say that latex film-forming is complete. In some cases crosslinking will also take place between, as well as within, the original particles, further enhancing the strength.

6.2.3 Effect of shrinkage and uneven pick-up

An associated effect of film-forming is that of shrinkage. The latex will typically be dipped at between 35 and 50% total solids content; the rest is water, most of which will be lost. As a result of this reduction in volume the film shrinks, which brings in its own set of problems. The interaction between the film and the former means that the film will be constrained to some extent, and therefore the shrinkage will result in a build-up of stress in the film. But as the particles integrate, the film becomes stronger, and in a film-forming polymer the build-up of strength is faster than the build-up of stress; in such a situation, a good film is obtained. If the polymer is not a film-forming polymer, then the resultant film will show the familiar 'mud cracking' as the stress caused by shrinkage will exceed any build-up of strength within the newly forming film. In an elastomer with a low wet-gel strength this shrinkage can lead to splitting of the film. Should this happen, the process will need to be modified to control the rate of drying to try and build strength faster, or the formulation modified to increase the wet-gel strength. Shrinkage can also cause the film to slide down the former, especially in coagulant dipping where it is common practice to include powders in the coagulant to act as a parting aid and help release of the rubber film from the former. Where the former is patterned this shrinkage can lead to a multiple imaging of the pattern on the finished product, and again action will have to be taken to remedy the situation, usually by modifying the drying and leaching conditions.

Another consequence of the shrinkage can be evident on straight dipped products such as condoms. Where there are sudden variations of thickness (e.g., where a drop forms on the end of the condom) then the rate of drying will differ between the thicker zone and the thinner zone. Thus, the rate of shrinkage and the build-up of strength will be different, and the thinner area will dry and shrink faster. The thicker, wetter area may not have sufficient strength to resist the shrinkage, and as a result a crack can form where the two zones meet. In extreme cases this crack can go all the way through the product. This is illustrated schematically in Figure 6.3.

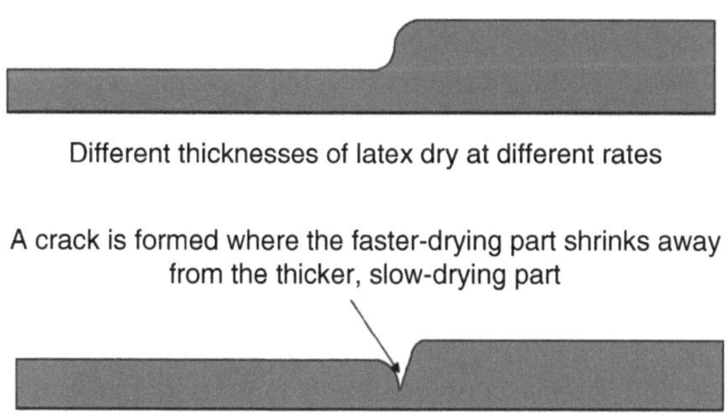

Figure 6.3: Thickness variations can lead to cracking. As the latex dries and shrinks, the thinner, drier area can pull away from the thicker, wetter area leaving a crack in the film.

A photograph of cracking on the end of a NR condom is shown in Figure 6.4. In this instance, the cracking has actually gone right through the film in places.

Figure 6.5 shows more general cracking on an NR latex film.

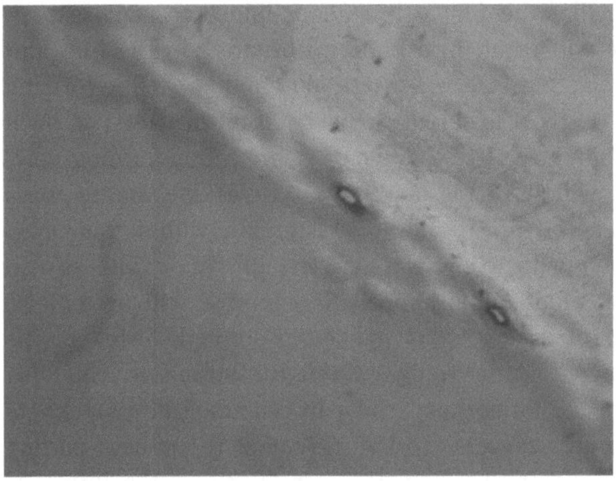

Figure 6.4: Cracking on the end of a condom.

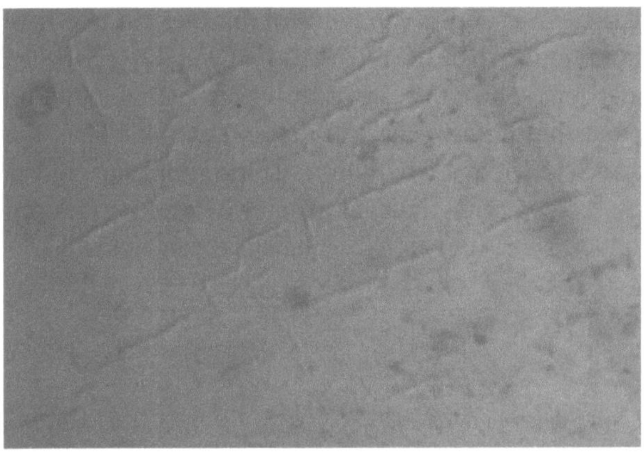

Figure 6.5: Cracking on a NR film.

These variations in thickness and subsequent cracking, can be a particular problem on the end of condoms or on the fingertips of gloves, where the faster withdrawal rate resulting from the curvature of the former can lead to an increase in the pick-up of latex or coagulant. At times, the cracking on the teat of a condom can be so severe that the teat can become detached. On gloves (and also on condoms) these thickness variations can lead to 'cupping', where the rubber film at the tip of the finger or the end of the condom inverts to form a shallow cup.

6.2.4 'Orange peel'

Straight dipped films can sometimes take on the appearance of 'orange peel' – the surface of the film is dimpled, just like the skin of an orange. This appearance is the result of the 'Marangoni' effect, and the individual dimples are caused by Bénard cells. As the latex dries, and water is removed from the surface of the film, the surface tension of the liquid at the surface increases. This difference in surface tension sets up a circulation of the remaining liquid, usually in very small cells. As the latex dries, these cells give the surface of the film the appearance of orange peel. This situation normally occurs when the film is even and the drying environment is stable. Thus, the appearance of 'orange peel' on the finished product is usually to be welcomed as the sign of a good film and a well-controlled dipping and drying process.

References

1. D.C. Blackley in *Polymer Latices: Science and Technology*, Volume 1, Chapman & Hall, London, UK, 1997.
2. D.H. Everett in *Basic Principles of Colloid Science*, Royal Society of Chemistry Paperbacks, London, UK, 1988.
3. ISO 35:2004: Natural rubber latex concentrate – Determination of mechanical stability.
4. ISO 2006-1:2009: Rubber latex, synthetic – Determination of mechanical stability – Part 1: High-speed method.
5. ISO 2006-2:2009: Rubber latex, synthetic – Determination of mechanical stability– Part 2: Moderate-speed method under load.
6. P.A. Steward, J. Hearn and M.C. Wilkinson, *Advances in Colloid and Interfacial Science*, 2000, **86**, 195.
7. T.D. Pendle and A.D.T. Gorton in *Proceedings of the International Rubber Technology Conference 1988*, Eds., L.L. Amin and L.K. Thong, Penang, Malaysia, 1989.

7 Latex dipping

7.1 Introduction

The usual process by which the required article is formed from a latex is known as dipping. A shaped former (also known as a 'form' or 'mould'), pretreated if necessary, is immersed in the latex, becomes coated with the latex and is withdrawn. The latex film is dried, the process repeated if required, and after drying the formed article is removed. The procedure sounds simple, and in general is; the challenge comes in being able to repeat the process on an industrial scale, making millions and millions of units of high quality, and at high yield and a viable cost.

Dipping can be carried out using many elastomeric latices, both natural and synthetic, but in order to yield useful product the elastomer must be film-forming, have sufficient gel strength to maintain a coherent film as it dries and have sufficient strength when dry for the end use. In practice, this means that latex dipping is generally confined to latices of natural rubber (NR), polychloroprene (CR), acrylonitrile-butadiene rubber (NBR) and synthetic polyisoprene rubber. Some artificial latices, such as synthetic *cis*-1,4-polyisoprene can also be dipped. Latices of styrene-butadiene copolymers, polyurethanes, polyvinyl acetate and various acrylic polymers and copolymers can be dipped, but as yet have little or no industrial applications.

After the product has been dipped it is removed from the former, but will generally need further treatments before the product is in the final form. Different post-treatments for dipped latex products are discussed in Chapter 11.

There are several ways by which the latex coating can be applied to the former, including:
- Straight dipping
- Coagulant dipping
- Heat-sensitised dipping
- Electrodeposition
- Spray coating

Of these, the first two are of major industrial importance, the third is used in some specialised applications, but the fourth, electrodeposition, is at present of little commercial significance. Spray coating using latices is still very much a process of the future and, at the time of writing, is confined to very small-scale or experimental work only.

These processes will now be considered in more detail.

7.2 Straight dipping

In straight dipping, the former is immersed into the latex and slowly withdrawn, and is the usual way in which condoms and many examination gloves

https://doi.org/10.1515/9783110638097-007

are produced. A film of latex will cling to the former by virtue of its viscosity. For straight dipping to be successful, the former must be wet easily and evenly by the latex. Wetting is covered in more detail in Chapter 8, and as described there, is a way in which the surface free energy of the system is reduced when a high-energy solid surface (the former) is covered by a lower surface energy liquid (the latex). For good wetting to occur, not only must the surface of the former be of the right material, it must also be clean and dry. Any surface contamination will change the surface energy of the former and result in poor wetting.

Straight dipping usually consists of the following stages: entry into the latex, exit from the latex (withdrawal), drying, beading, vulcanisation, leaching, product stripping and former cleaning. In many cases there will be two, or even more, latex dips, each with its own drying stage. A schematic of straight dipping is shown in Figure 7.1, and each stage is discussed in more detail below.

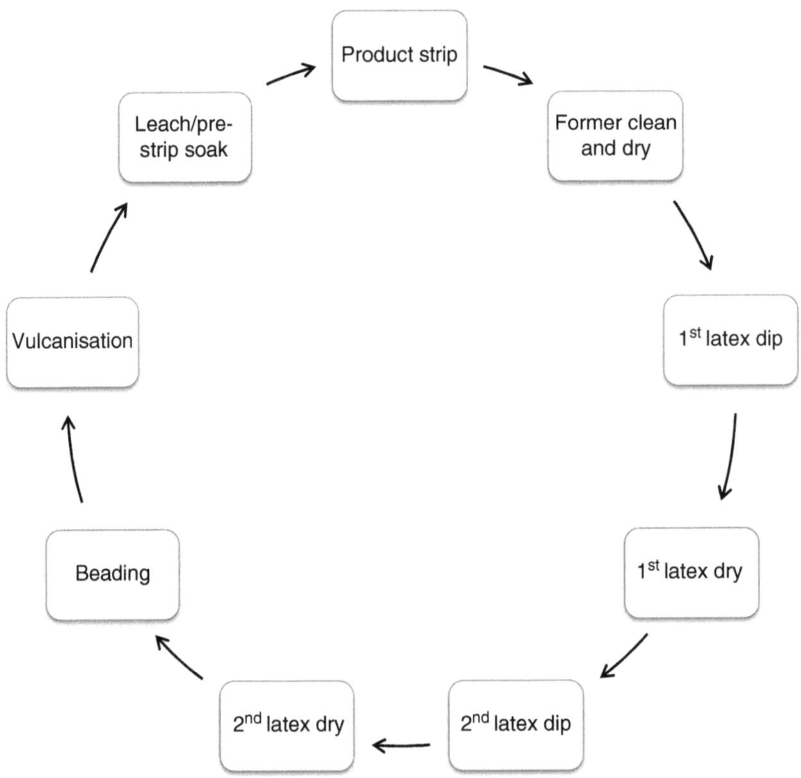

Figure 7.1: A schematic diagram of the straight dipping processes.

7.2.1 Former entry into the latex

The entry speed of the former into the latex in straight dipping is generally not a major design criterion. The main potential problem is that air will be entrained into the latex if the entry speed is faster than the speed with which the latex wets the former, and air bubbles in the latex are a potential source of defects and best avoided. As noted in Chapter 8, the speed of wetting of formers can be calculated, for example, by work done on the coating of magnetic tape media [1]. In this study the critical entry speed, 'V_{ae}', is defined as shown in Equation (7.1):

$$V_{ae} \approx 70.5\left[\mu(G/\rho\sigma)^{0.5}\right]^{-0.77} \tag{7.1}$$

where 'μ' is the viscosity of the liquid, 'ρ' is the density, 'σ' is the surface tension and 'G' is the gravitational constant.

If the former enters the latex faster than this speed of wetting, air will be pulled into the latex. However, with normal latex formulations, typical line speeds and glass or porcelain formers, air entrainment on entry is not normally a problem. The entry speed must also take account of the former design. When there are any undercuts, or other potential places where air can be trapped, for example, the crotch between the fingers of a glove former, then the entry speed into the latex may have to be reduced. In practice, observation of the meniscus at the interface between the latex and the former at the entry into the latex can yield valuable information about the cleanliness of the former and the evenness of the wetting.

If the surface of the latex dips down erratically as the former enters the latex and seems reluctant to wet the former surface, then it is likely that the surface of the latex is not clean enough. If the latex seems to 'climb up' the former as it enters the latex, then it is likely that the surface of the former is damp. Formers that are slightly damp will not normally be a problem, but this situation is best avoided if possible. Wet formers may not always give rise to any film defects, but water droplets on the former are an uncontrolled addition of water into the latex, and may cause the viscosity and total solids content of the latex to drift out of specification. A clean former will wet smoothly and evenly.

7.2.2 Former exit from the latex

The exit speed in straight dipping is important, as this is one of the primary ways of controlling product thickness. Somewhat counterintuitively, the faster the former is withdrawn from the latex, the greater the amount of latex that coats the former (known as the 'pick-up') for a given latex viscosity. Conversely, the slower the exit speed, the lower the pick-up. The pick-up is controlled by the rate at which the latex drains down the former, hence the faster the former is removed from the latex, the less time the latex has to flow off it. This flow is the reason why so many straight dipped articles are thicker towards the closed end, compared with the open end. The latex at the top of the article (as dipped) has more time to flow down the former than that dipped at the bottom, hence more latex will collect towards the closed end of the former. If required, the initial stages of the withdrawal can be made more rapid than the later stages [2], so that the top part of the product picks up more latex than the lower parts. This latex can subsequently drain down and help to give the product a more even thickness. Note that with product dipped on a chain plant, thickness variation can occur around the circumference of the product, as well as along the product. This is thought to be caused by the turbulence induced as the formers pass through the latex, as the formers will virtually always be moving faster than the latex is being circulated.

Sometimes the formers will be rotated whilst they are in the latex, in an attempt to even out the film distribution. However, rotating the formers on exit (or on entry) increases the effective withdrawal (or entry) speed, as now the surface of the former will leave (or enter) the latex in a more rapid spiral motion, rather than a linear motion. On entry, this will increase the risk of air entrainment and on exit will increase the pick-up proportionally. Extra latex picked up may increase the propensity for uneven latex distribution, and the problems associated with it, such as poor drying and film cracking. It is usual to move the formers up into an approximately horizontal position, or even above the horizontal, and rotate them after they have left the latex, to help distribute the latex evenly and prevent any further flow down the former as the latex film is drying. It is advisable that this movement takes place as soon as possible after the exit from the latex, to minimise the flow of latex down the former, but note that the former *must* have left the latex surface. Swinging the former up too soon, before the former has left the latex, will leave a heavy deposit at the tip of the former, brought about by the sudden increase in exit speed.

As the very tip of the former leaves the latex, it is not uncommon for a bubble to form as the film between the former and the latex surface breaks. This is thought to be caused by a 'neck' of latex being formed, albeit briefly, between the surface of the dip tank and the former as the former leaves the latex. When this 'neck' breaks, a bubble can be created. These bubbles can remain on the surface of the latex, on the film picked up by the former or both. Those picked up by the former will often burst as soon as they enter the drying oven, but sometimes they persist and cause a potential defect in the finished product. A gentle jet of air directed at the tip of the former as soon as it has left the latex can be used to burst any bubbles before the film starts to dry, but the best way to remove them is to change the rheology of the latex and the dipping conditions slightly, for example, a small increase in exit speed, with the appropriate reduction in latex viscosity, or a reduction in speed/increase in viscosity will often remove the problem.

7.2.3 Drying

After the latex dip, the formers will pass to the drying stage. On a chain plant this usually takes the form of a heated tunnel through which the formers pass, rotating as they go. On a batch plant the oven will be a heated chamber. The heating can be radiant heaters, hot air circulation or a mixture of both. This drying stage is often assumed to be simple, but the opposite is the case. The design of the oven is important, as it is preferred to have a small enclosure to minimise heating costs, but the enclosure must be large enough to accommodate the number of formers dictated by the production speed. Drying at a high temperature for a short time means that the ovens can be small, but these conditions can cause other problems such as rapid skinning over of the film or excessive drying. The entry and exit to the ovens on a chain plant must be wide enough to allow any formers that are slightly misaligned to pass through without hitting the sides, but small enough to minimise heat losses. It is also common practice on a chain plant to have different zones within the drying tunnel that can be maintained at different temperatures, to control the rate of drying, and again the entry and exit apertures must be correctly sized. It can also be advantageous to have individually switched heaters positioned in these heating zones to cover the different zones of the former – bead area, central area, tip and so on. By controlling which heaters are switched on at any one time, the heat can be focused to ensure that the drying of the latex film is even, and help to avoid any film cracking. The heaters can be positioned at the bottom, top or sides of the ovens, but note that if they are on the bottom of the oven, latex may drip off the formers onto the heaters, especially if, on a chain plant, the line is stopped for any reason, giving rise to smoke and the smell of burning rubber. Batch plants can utilise programmed ovens to ramp-up the temperature and control the rate and efficiency of drying. A cooling stage or zone can be used after the drying, to cool the former before any subsequent operations. Note that the latex in the dip tank can have a total solids content of 40% or even lower – in other words, a large part of the mass of the latex picked up is water and will be lost as it dries causing the film to shrink onto the former. As discussed in Section 6.2.3, this can have consequences with regard to cracking if the wet-gel strength of the elastomer is not high enough, and can also make stripping more difficult. A drying process that is as gentle and progressive as production constraints allow should be used. If the drying is aggressive, for example, if the drying time is short and temperatures high, or the pick-up is heavy, then the latex film will tend to adhere more strongly to the former and stripping will become more difficult. In fact, one of the commonest causes of difficulty in stripping straight dipped articles is overdrying somewhere on the dipping line, and one of the commonest solutions to this problem is to ease off on the drying temperatures. Aggressive drying can also lead to overdrying bubbles, as these drying conditions can often mean that the outside of the latex film becomes dry whilst the inside is still wet. If the temperature is too high, then the water in the underlying wet layer can vaporise, forming bubbles that can subsequently lead to holes and product failure. Such defects usually occur in any thicker areas caused by uneven latex pick-up or flow, which is one of the reasons why such situations are to be avoided. The temperature of the oven can be controlled by thermocouples or, more sophisticatedly, by infrared monitoring of the film temperature. Typically, the temperatures in these drying ovens will be in the range of 60 to 100 °C, although it may be higher for some materials such as CR latex.

7.2.4 Subsequent dips

A second, or even a third dip can be used, depending on the product and its intended application. Subsequent dips will generally follow the same procedures as the first, although note that these subsequent dips are coating a rubber surface, rather than the former material. This will usually be a lower energy surface, hence any deficiencies in wetting may show up more here. Owing to the relative ease with which the former surface can be wet, the first dip is often slightly heavier than subsequent dips, whether by accident or design. This 'split weight' can be as much as 55% of the finished product on the first dip in a two-dip process and 40% in a three-dip process, although figures of around 52 and 35%, respectively, are more common. If the former is heavily patterned, too heavy a pick-up on the first dip will lead to drying problems, cracking and possible air entrainment. In this case, the solids content/viscosity of the first dip can be reduced to lower the pick-up and the product weight compensated in subsequent dip(s), although having different latex compositions in the different dips and their supply reservoirs will increase manufacturing complexity.

Note that the temperature of the former as it exits a drying stage and enters the next latex dip is important. If the former is too hot it can induce instability in the latex and give rise to coagulum, or initiate some uncontrolled heat-sensitised dipping. Hot formers can also accelerate the formation of a skin on the latex surface, even in dip tanks that are fitted with cooling jackets or coils. Any temperature variations or gradients that are created in the latex will affect the latex viscosity and hence pick-up, as well as sedimentation of any of the solid ingredients. In general, the temperature of the film going into the latex dip should be as low as practical, preferably below 40 °C. To help this, a cooling stage is often added at the exit of the intermediate driers. The degree of drying of the latex film during any subsequent dips is important: if the film is still damp (usually giving the film a somewhat cloudy appearance) when it enters a subsequent latex dip, then uneven drying will result, which can lead to film cracking. A damp film is much weaker than a dry film and is therefore more prone to damage. Conversely, if the film is too dry then subsequent layers may not bond well and the two films can delaminate. Drying conditions should be adjusted so that the rubber film has just lost any cloudy appearance by the time a subsequent dip is applied.

7.2.5 Beading

The next stage after the rubber film has been formed in straight dipping will usually be beading. The top edge of the dried film will be rolled down to give a strengthening rim to the product. Although it can appear to be simple, beading is a critical part of the production process, and beading defects are usually one of the major defect categories. Although the beading can be performed by hand in low-volume processes, more usually the rolling is carried out by a rotating brush or roller, or combination of brush and roller, which can be set at a slight angle to the axis of the former. Sometimes flexible rubber flaps mounted on a cylinder are used. Brushes are generally preferred for the first stage in the beading, as the edge of the latex film can be 'picked up' more easily by a brush. However, the type of brush – the diameter, bristle length, bristle stiffness and bristle material, for example – can be critical and may need to be optimised by trial for each type of product, dipping plant and line speed. These brushes will wear, especially the first ones after the prebead oven, which are exposed to the most heat and have to do the most work. It is therefore common practice on a chain plant to construct the brushes in segments, so that a worn segment can be replaced as required without having to replace the whole brush. An example of a beading brush on a condom dipping line is shown in Figure 7.2.

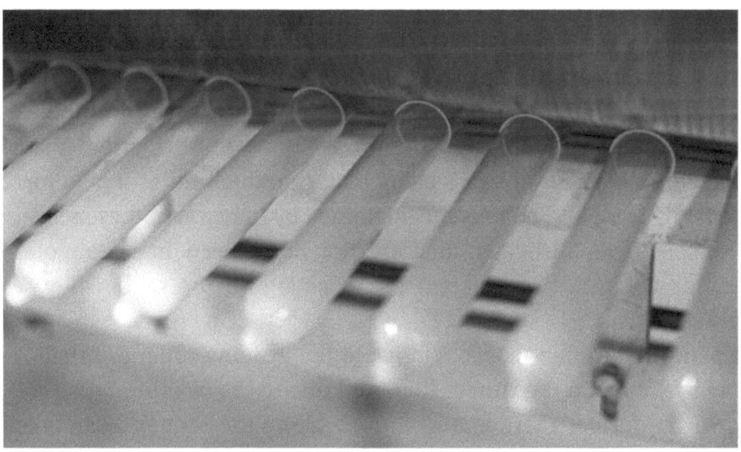

Figure 7.2: Beading brush on a condom dipping line. Reproduced with permission from the Thai Nippon Rubber Industry Public Company Limited. ©Thai Nippon Rubber Industry Public Company Limited.

For beading to be successful, it is important that the rubber film is in the correct condition to be beaded. It needs to be dry enough to have sufficient strength to be beaded, but not so dry that any inherent level of tack is reduced. As mentioned earlier, an overdried film is difficult to release from the former. Excessive drying can result in the beading brush or roller not picking up the edge of the rubber film immediately, and when it is picked up it will be uneven, giving rise to a lump or 'nodule' on an otherwise smooth bead. If the drying is excessive, or if the extent of prevulcanisation is too great, the beading can also be very rough or 'serrated'.

Hence, the dryness of the film is very important: it should be dry, but not too dry. A film that still contains water may be too weak to withstand the stresses involved in beading, and furthermore, any water trapped as the bead is formed may cause the bead to blister and 'blow' if too much heat is applied in subsequent operations. In a multidip process, it is common practice to make the first dip slightly longer than subsequent dips, so that there is a thinner film at the top of the article, which is easier for the brush or roller to pick-up and start the beading process. One of the potential problems with using a three-dip process is that the first dip may be too dry by the time the former comes to the beading station. If this occurs, consider altering the dip levels so that one of the subsequent dips provides the initial bead pick-up. Successful beading relies on the rubber film having sufficient tack to adhere to itself and form a coherent bead, which means that articles made from some synthetic latices, such as the medium- and high-acrylonitrile (ACN) grades of NBR latices cannot be beaded.

As the bead is rolled down, the formers will be rotated. On a batch plant this can be through a drive that can be coupled to the formers. On a chain plant this rotation can be driven by the motion of the chain rolling the formers over a contact strip. Alternatively, the formers can be driven by a rotating belt and very often both mechanisms are used, so the former is rotated, then counter rotated during bead formation. The reason for this is to avoid the problem known as 'figure-8 beads' (Figure 7.3), which are the result of stress being built into the bead as it is formed. When the article is stripped, and no longer constrained by the former, these built-in stresses distort the bead. Simplistically, the formation of the bead on a chain plant can be described as the result of different forces acting on the rubber film, as shown in Figure 7.4.

On a chain plant, the former moves relative to the beading roller or brush resulting in a sideways force on the rubber film, and due to the beading brush rolling the film down the former there is also a force along the former axis. Taken together, these forces result in an angled force on the bead. When the article is removed from the support of the former, this force relaxes, and can distort the bead so that it resembles a figure-of-eight. By changing the direction of rotation during the beading process, the stresses imposed by the movement of the former are reversed, and if done correctly will be cancelled out. Sometimes, only the rotation of the former will be reversed during the beading operation – there will be no rotation caused by the former movement along the dipping line – and this can be sufficient to give good beads. The problem with figure-8 beads is much less

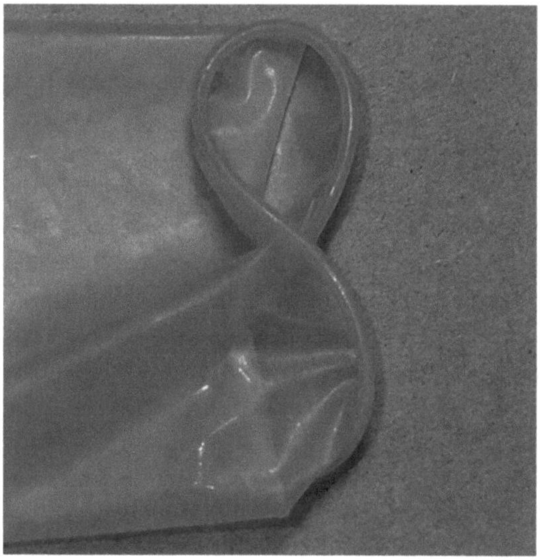

Figure 7.3: A 'figure-8' bead on a condom.

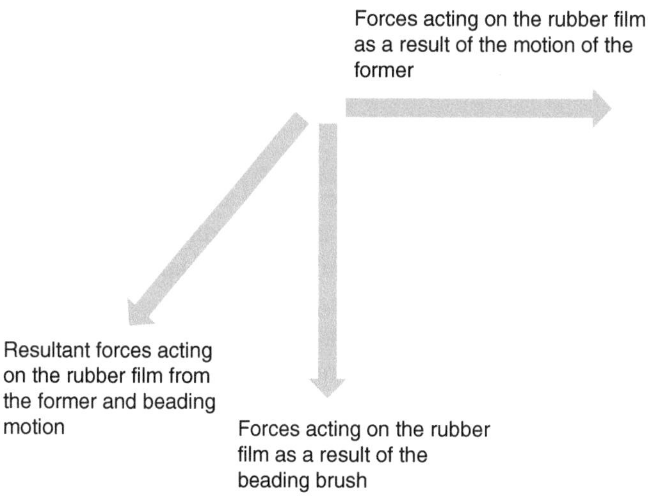

Figure 7.4: A simple representation of the forces acting on the product during beading on a chain plant.

frequent on a batch plant. As will be realised, having a beading station that allows a good range of adjustments, of beading brush or roller position, angle, and preferably control over the rotational speed of the former and brush/roller will ensure that good beads can be obtained under a range of different production conditions.

7.2.6 Vulcanisation

As many NR latex dipping lines use a fully or partially-vulcanised latex, the term ‘vulcanisation’ for the final heating stage in these cases is somewhat of a misnomer, although in wide usage. In addition, the constraints of heating the film on the production plant, for example, limited time, the need to avoid overheating the film and causing stripping problems and blisters, and the energy needed to heat a former and spinner assembly as well, mean that any vulcanisation is generally better employed in off-line drying ovens. Synthetic latices are often not prevulcanised (although a period of maturation may be employed), and in this case a final heating stage is necessary to give sufficient strength to the rubber film to allow subsequent stripping and processing. The design of the vulcanising oven will often be very similar to the other drying ovens – in fact, on a batch dip plant they may even be the same oven. On a chain plant there may be several different zones, allowing a graded temperature profile to be used, so that all the moisture is removed from the film before the temperature is raised above 100 °C. Batch plants can use programmable ovens to ramp up the temperature gradually and obtain the same effect. The temperatures used in vulcanisation can vary greatly depending on the product: a prevulcanised NR product may only be heated to 40 or 50 °C, whilst a CR product may need temperatures of 125 °C or more.

7.2.7 Leaching

Leaching is usually taken to mean immersion of the dipped article, whilst still on the former, into a bath of warm or hot water. With a straight dipping process the leaching stage almost always takes place after vulcanisation. The purpose of leaching is two-fold: to remove any water-soluble materials and to loosen the product from the former ready for stripping (a ‘prestrip soak’). Efficient removal of water-soluble materials requires sufficient removal and replenishment of the water to maintain extraction efficiency, a situation that is rarely encountered in practice, owing to the energy costs of continually discarding and renewing hot water. As a result, most leaching baths on dipping plants contain a relatively high level of extracted material. The efficiency can be improved by using a second separate leaching stage. In practice, the bulk of extractable material is probably removed during any washing stages in post-treatment, where the washing times are sufficiently long and the washing liquid may be changed or replenished more often. Studies on leaching of dried rubber films dipped from prevulcanised NR latex by Gazely [3] have shown that the extraction of water-soluble materials from such films is faster than from a similar film which has been postvulcanised. The rate of extraction is also affected by the degree of prevulcanisation – the more highly prevulcanised the latex is, the faster these materials can be extracted. This is attributed to the lower degree of interparticle coalescence in a film dipped from a prevulcanised latex, allowing the water to penetrate the rubber film and dissolve any water-soluble materials more easily. Gazely also found that the rate of extraction of materials from the rubber film decreased as the drying temperature increased, which is assumed to be the result of the higher drying temperatures promoting greater interparticle coalescence. A higher leach temperature also increases the rate at which any extraction of material occurs. Removal of these water-soluble materials has also been found to increase physical properties, especially tensile strength and modulus [4, 5], and such increases can, at times, be substantial. It is thought that the presence of these leachable materials (e.g., residual soaps, proteins and so on) interferes with the interparticle coalescence. Removing them allows the latex particles to integrate better and the rubber film to become much more coherent, and thus stronger. Other benefits from an effective leaching procedure are the reduction of any potential human allergens, such as accelerator residues or extractable latex proteins, and a reduced sensitivity to water. Extensive and effective leaching is thus highly recommended to ensure a product is safe and fit for purpose. Good leaching can also lead to improved clarity in the finished dipped article.

The second function of the leaching process is to loosen the rubber film from the former, which is most easily accomplished by using an alkaline solution as the leach. The rubber particles still retain some of their anionic character, and this means that the film can be 'opened up' by an alkaline solution. Alkalis will also help solubilise any anionic soap residues and aid in their extraction. The alkali most commonly used is sodium hydroxide (NaOH), but other alkalis such as potassium hydroxide or even sodium carbonate (Na_2CO_3) or potassium carbonate can be used. A concentration of 0.5 to 1.0% of NaOH or 2 to 3% of Na_2CO_3 is usually used, although the concentration does not seem to be important, as easy and effective stripping can be achieved with very low levels of alkali in the leach. Another material sometimes used in the leach water is sodium triphosphate (STP), also known as sodium tripolyphosphate. STP acts as a water softener and chelates heavy metal ions, as well as helping to loosen the film from the former. Unfortunately, STP is also a eutrophic material (that is, it acts as a nutrient for some plant life and promotes the growth of algae and so on, in local watercourses and lakes), and as such has given rise to concerns regarding its environmental effects. As a result, its use in prohibited in some countries. Ammonia is highly effective at penetrating the film and releasing the rubber from the former, although its volatility means that it will not persist long in hot leach water and unless there is efficient extraction over the leach tank the ammonia vapour will contaminate the air in the vicinity of the leach. However, it is a very useful 'troubleshooting' measure in the event of stripping problems: adding a litre or two of ammonia into the leach tank can have an almost miraculous effect on products which are difficult to strip. It is easy to see how effective a leach solution is at penetrating a film, as a fully hydrated film will appear white. A leach tank will usually be maintained at a temperature of 40 to 70 °C.

Temperature does not appear to have much effect on the way in which the leach loosens the film, although as described above, a higher temperature leach is a more effective extraction medium.

7.2.8 Stripping

Stripping can be carried out manually, but in general, straight dipped products will be stripped by some automatic process. For condoms, this can be by high-pressure water jets, rotating brushes or rotating flexible rubber flaps, whilst examination gloves are often stripped by water jets. A jet of compressed air may be used as well, to inflate the product partially and help release it from the former. Dry stripping of products, where powder and, for example, silicone emulsion, are applied to the surface of the dried product before stripping to counteract the tack before removing from the former, is sometimes used. However, this procedure carries considerable risks of contamination of the formers and latex, and therefore virtually all high-output dipping lines now use a wet-stripping process. Water-jet stripping can be carried out either underwater or above water, using two or more high-pressure water jets directed onto the former. On a chain plant, successive jets can be aimed progressively further down the former to ensure that the product is stripped completely. In particular, there needs to be a jet aimed directly at the lower end of the former, as this is where the rubber film tends to be thicker and the product will cling onto the former more tightly. The stripping solution will either be soft water or a dilute alkali solution. Dilute solutions of surfactant can also be used, but it is recommended that a non- foaming surfactant is used, otherwise unacceptable levels of foam will be generated around the stripping area. Foaming can also be a problem with an alkaline-stripping solution, when there are residual fatty acid soaps in the rubber film that are extracted by the stripping solution. In some cases, a small amount of powder will be added to the stripping solution to prevent the products sticking, and it is usual practice to recirculate the stripping solution. Condoms in particular can also be stripped from a chain plant using rotating brushes, one on each side of the former and angled so as to wipe the product down the former. As indicated above, it is important to ensure that the brushes continue to the end of the former to completely remove any product. Any partially stripped product that remains on the former and is carried into the latex dipper will contaminate the latex with stripping and washing fluids. The stripping brushes should not be too tight, otherwise the product may be damaged. An easy way to check the tightness of the brushes is to try and rotate the former whilst it is in the brushes: if the brushes are adjusted correctly it will be possible to spin the former with only a small amount of resistance. One of the problems with both water-jet and brush stripping is that the product may be turned inside out during the stripping. For a standard condom or an ambidextrous glove this will not be a problem, but if the orientation is important – for example, if the gloves are made as the left or right hand, or if a coating is applied preferentially to one side – then the stripping must be designed and adjusted so that product inversion either does not occur, or always occurs. If this cannot be guaranteed, then another way must be found to strip the product from the former.

After stripping, the product is collected underneath the stripping zone or may be transported away from the dipping line. In either case, it is important that the product is kept wet until subsequent processing operations. If the products are allowed to dry out, then they will stick together and may not separate during subsequent processing. Sticking can also be a problem if those products at the bottom are compressed for a long time by the weight of products being added above them, and can impose constraints on batch collection times and size.

7.2.9 Former clean

The last stage in a straight dipping process – or the first stage, depending on the perspective – is former cleaning. Clean, dry formers are essential for maintaining good product quality. It is common practice to rinse off the formers with soft water immediately after stripping and then wash with soft water, often accompanied by a scrub with brushes. The water may be cold or warm. After washing the formers must be dried. Drying can be carried out with heat, with more brushes, with jets of compressed air, or with any combination of these methods. On a chain plant there is often a brush, sponge or towel positioned to wipe over the very end of the former to remove any final drops of water. It is essential that these brushes and so on, are maintained in a clean condition, to prevent contamination of the formers. If the formers are slightly damp, it is not a serious problem, but is better avoided if possible. The impact of water being carried into the latex by wet formers should also not be discounted. For example, if each former on a chain dipping plant running at 75 pieces per minute carries 5 drops of water, then over 1 litre of water per hour will be carried into the latex. An uncontrolled addition of water at this level can cause problems in maintaining the product weight. Despite these routine cleaning procedures, there will be times when a more thorough cleaning is required. This may be undertaken when there are signs of problems such as poor wetting, or, preferably, as part of a regular preventive maintenance programme. More thorough cleaning usually involves washing the formers with dilute alkali or acid. Dipping is often suspended whilst former cleaning is in progress, although if sufficient precautions are taken to ensure that all cleaning fluids are removed from the formers, the production process can continue as normal whilst the formers are being cleaned. A typical cleaning solution is dilute – say 5% – citric acid, which is preferred to other acids such as hydrochloric or sulfuric acid, as it is cheap, available in a high state of purity and is much less corrosive to the production plant. Cleaning with alkalis is effective, but note that both glass (especially soda glass) and porcelain can be affected by alkalis, and the surface may be etched away by alkalis over time. Figure 7.5 shows a soda glass condom former where the surface has been etched by cleaning over time.

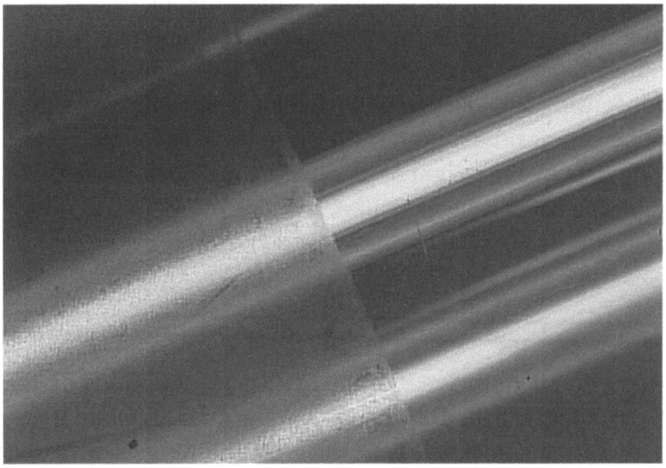

Figure 7.5: A soda glass condom former, showing how the surface has been etched by cleaning over time.

7.3 Coagulant dipping

Whilst straight dipping relies on the viscosity of the latex and the former withdrawal speed to build the required product thickness, coagulant dipping uses a destabilising chemical (the 'coagulant', sometimes also called the 'coacervant') coated onto the former. When the former is immersed into the latex, the coagulant will react with the colloid stabiliser of the latex in the immediate vicinity of the former. The stabiliser is destroyed, the rubber particles coalesce and the rubber forms a gel around the former.

As the coagulant diffuses further into the latex more and more of the latex will form a gelled layer of rubber on the former, which can then be withdrawn and dried. The thickness of the latex film picked up will depend on the type and concentration of the coagulant, the stability of the latex to the coagulant and the length of time that the former remains in the latex (the 'dwell time'). Using coagulant dipping will allow a greater pick-up, and hence a thicker product than straight dipping, and as such is usually the method of choice for gloves, balloons, catheters and so on. With the correct choice of coagulant type, concentration, dwell time and latex total solids concentration, films of up to approximately 1 mm can be obtained in a single latex dip. Many of the steps in coagulant dipping are similar to those in straight dipping, and a schematic diagram of the production sequence is shown in Figure 7.6. The principal stages in the process are described below.

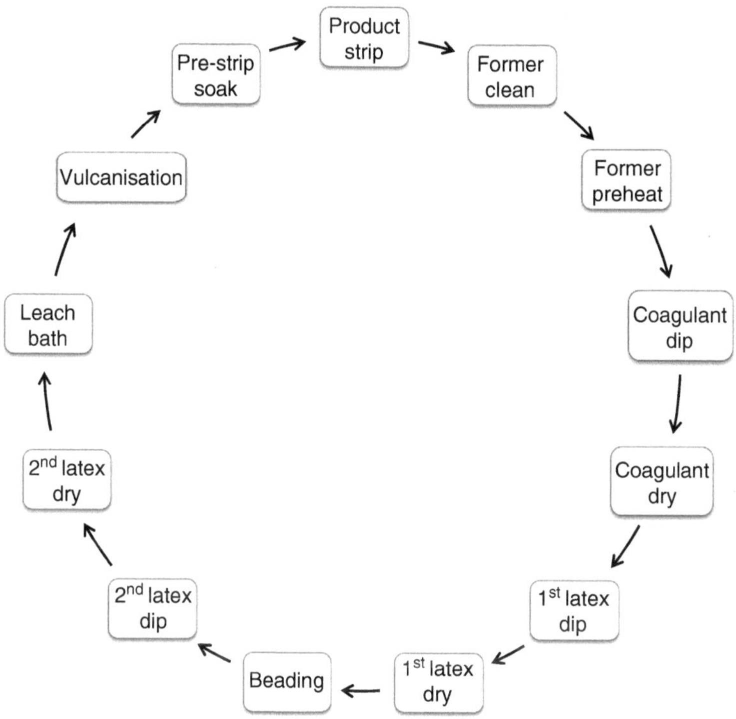

Figure 7.6: A schematic diagram of the processes involved in coagulant dipping.

7.3.1 Former preheat

It is necessary to preheat the former after cleaning, to assist in the drying of the coagulant. This preheating is usually carried out by immersing the former in a tank of hot water, although an oven may be used. As coagulant dipping is generally the method of choice for gloves, and glove formers can be quite large and have a high heat capacity, the more rapid heat transfer effected by hot water is preferred. Soft water should be used, as hard water will create a scum when heated and leave a deposit on the former, which will affect the quality of the latex film. The temperature of the preheat tank will depend upon the process, but will typically be around 70 °C.

7.3.2 Coagulant dip

Any material that can destabilise the latex can, in principle, be used as a coagulant in coagulant dipping, and include any di- or trivalent metal salt, acids or alcohols. However, there are several constraining requirements for the ideal coagulant chemical. It must be soluble or miscible in water or alcohol, and the coagulant solution must coat the former evenly and be retained on the former. The reaction with the latex stabiliser system must be rapid and progressive. The resultant gelled film must be firm and coherent and the coagulant must be safe to use. Any coagulant residues in the latex film should be easily removed, and not detract from the performance and safety of the finished article. In practice, these constraints mean that there are a limited number of materials used as coagulants in commercial practice. The following are amongst the ones most commonly encountered:
- Calcium chloride ($CaCl_2$)
- Calcium nitrate [$Ca(NO_3)_2$]
- Acetic acid
- Cyclohexylammonium acetate

The use of these coagulants gives a firm, coherent gelled film from the latex, although acetic acid and cyclohexylammonium acetate will leave a residual odour of acetic acid. This odour will usually dissipate over a period of time, and can also be removed by thorough washing. In practice, most coagulant dipping will use solutions of $Ca(NO_3)_2$ or $CaCl_2$. These materials are readily available, cheap and give a good gelled rubber film which can withstand subsequent processes on the dipping line. These calcium salts are deliquescent, hence films from their solutions do not dry out completely on the former under the drying conditions normally encountered. This viscous fluid film of coagulant appears to be more effective at coagulating the latex film. Cyclohexylammonium acetate is often used when dipping supported gloves, probably because its milder action allows a better bond to be formed between the fabric liner and the gelled latex, whilst still preventing 'strike through' of the latex through the fabric. Cyclohexylammonium acetate can also be used when the clarity of the finished product is important, as the products of the reaction between it and commonly used latex stabilisers, such as fatty acid soaps, are more soluble in the dried rubber film than the corresponding calcium salts formed when $Ca(NO_3)_2$ or $CaCl_2$ are used. Thus, the finished film will contain fewer visible solid particles.

The coagulant material will be dissolved in water or alcohol – usually ethyl alcohol, in the form of industrial methylated spirits. Although material costs will be higher than with an aqueous solution, an alcoholic solution can wet the former more evenly and will dry faster and thus can offer savings in time and energy, although the flammable nature of alcohol will necessitate special handling precautions. An alcohol–water mixture can also be used, which reduces flammability whilst retaining many of the benefits of an alcoholic coagulant.

The formulation of the coagulant solution will vary greatly, depending on the dipping machinery being used, the application, the type of product and the latex being dipped. As well as the actual coagulant material itself, an aqueous coagulant solution will generally also contain a small amount of an inert powder, for example, finely divided calcium carbonate. The purpose of the powder is twofold: it will increase the viscosity of the coagulant solution as it dries, thus reducing the flow of the coagulant and helping retain an even film of coagulant on the former; and it will also act as a 'parting aid', helping to release the dried rubber film from the former. As a side observation, the addition of a little powdered sulfur to the coagulant can be a useful remedial measure if there are problems with a soft cure. A small amount of surfactant will also be used in an aqueous coagulant to help disperse the powder and ensure good wetting of the former. Non-ionic surfactants will usually be used, and those with a low tendency to foam are best. Surfactants that are foam promoters will cause webbing – the temporary formation of a film between the fingers of a glove former as it is withdrawn from the coagulant solution. Excessive webbing is to be avoided if possible, as when the web breaks it will leave an uneven coagulant film, which translates into an uneven latex film. If webbing is a problem, special antiwebbing agents can be added, such as medium chain alcohols or silicone emulsions. There are also several effective proprietary antiwebbing agents of undisclosed formulation available. Note that many non-ionic surfactants have a 'cloud point' at elevated temperatures – that is, they become less soluble when the temperature is increased. It is necessary to ensure that the chosen surfactant will be adequately soluble at the temperature of the coagulant bath. Four examples of typical coagulant bath formulations are shown in Table 7.1, although as previously stated, the exact formulation, especially the type and concentration of coagulant, will depend upon the product type and latex formulation.

Table 7.1: Examples of coagulant formulations for coagulant dipping.

Ingredient	Percentage by weight			
	A	B	C	D
Hydrated calcium nitrate	50	30	–	–
Anhydrous calcium chloride	–	–	30	–
Cyclohexylamine acetate	–	–	–	25
$CaCO_3$	1.5	1.5	1.5	–
Surfactant	0.25	–	0.25	–
Water	48.25	18.5	68.25	75
Industrial methylated spirits	–	50	–	–

Both alcoholic and aqueous coagulant solutions will change in concentration as the liquid is lost through evaporation. Control of the coagulant concentration will therefore be necessary and can easily be carried out by monitoring the specific gravity of the solution, using a suitable hydrometer. The correct coagulant concentration on the dipping line will be maintained by controlling the concentration in the feed reservoir. The temperature of an aqueous coagulant bath will typically be around 70 °C, although an alcoholic coagulant will be run at a much lower temperature of 30 to 35 °C – the rate of drying at this temperature is adequately fast and higher temperatures lead to rapid loss of the solvent through evaporation.

The formers are dipped into the coagulant and withdrawn. The entry and exit speeds are usually fast, and the pick-up of the coagulant solution on the former will depend upon the withdrawal speed (as in straight dipping), the type and roughness of the former surface, the viscosity of the coagulant solution and the speed with which the viscosity increases as the coagulant dries. The formers will usually be rotated and raised above the horizontal after they have left the coagulant dip to help give an even distribution of the coagulant as it dries, although this may not always be necessary with an alcoholic coagulant. The coagulant tank will require considerable agitation if the coagulant solution contains a powder parting aid, in order that the powder is kept in suspension.

7.3.3 Coagulant dry

The coagulant needs to be substantially dry before the former enters the latex dip. If the coagulant is still wet, the coagulant solution will wash off the former into the latex dip, forming coagulum in the latex dip tank. On the other hand, if the coagulant on the former is completely dry the reaction with the latex will not be as effective. As mentioned above, one of the advantages of $CaCl_2$ and $Ca(NO_3)_2$ is that they are deliquescent under the right conditions, and tend to give a viscous liquid film on the former, which is very effective at promoting a firm coherent gelled latex film. The coagulant drying stage is usually a radiant or hot air oven, although if a hot alcoholic coagulant is used, an oven may not be necessary. As with straight dipping, it is better if the former is not too hot entering the latex, as this can cause instability and other problems, as described earlier.

7.3.4 Latex dip

The coagulant-coated former will now be immersed into the latex tank to form the dipped article. Whilst the entry and exit for the preheat and coagulant tanks are usually simple 'in–out' affairs, the entry into the latex dip can require a more complex arrangement, especially for formers such as glove formers. In these cases, the ideal entry speed will be fairly rapid as the fingers enter the latex, and then slow down to avoid air being trapped in the crotches between the fingers and finger and thumb. The entry will then speed up again once the crotches are immersed in the latex. With gloves and other beaded products it is common practice to dip to full immersion and then immediately withdraw the former by some 5 to 10 mm. This gives a thin film at the open end of the product that is more easily picked up by the bead rollers. Special attention must be paid to the dipping of supported gloves, as the hydrostatic pressure deeper in the latex can help force the latex through the liner, creating 'strike through', which is undesirable, as it stiffens the glove, reducing flexibility. For this reason, many supported gloves will be dipped with the former at an angle, rather than vertical, so the depth of immersion into the latex is reduced. Long gloves may also be dipped at an angle to reduce the exit time and help control the evenness of the film.

The dwell time in the latex will depend upon the thickness required – the longer the dwell time, the thicker the film. Withdrawal of the former from the latex is usually moderately rapid, in order to minimise thickness variations along the dipped article. As with straight dipping, the speed of withdrawal may be varied during the exit in order to control the wall thickness over the length of the product. Note that there will also be some pick-up of uncoagulated latex through a straight dipping mechanism, and if this pick-up is excessive any latex drips falling back into the dip tank may contain coagulant and thus lead to coagulum in the latex. It is common practice to change the attitude of the formers to horizontal or higher after they have left the latex surface, and rotate them to even out the latex film as it dries.

The mechanisms which come into play as the coagulant-coated former enters a latex have been the subject of considerable research, and there is a wealth of detailed information in the literature. Whilst the actual mechanisms are complex, the net result is that the coagulant reacts with the stabilising system in the latex and reduces its effectiveness. Without effective stabilisation the rubber particles coalesce together and form a film of rubber over the former surface. The longer a coagulant-coated former remains in the latex, the thicker the rubber deposit that will be laid down; however, the process is not constant. The process of coagulation and deposition is initially rapid, but then tails off [6]. This phenomenon seems to be independent of coagulant type, coagulant concentration, latex type and latex solids content. It is thought to be caused by the fact that diffusion of the coagulant ions (usually Ca^{2+}) is impeded by the coalescence of the rubber particles as coagulation proceeds, and the depletion and dilution of the coagulating ions as they diffuse out from the former.

Nevertheless, experimental work [6] has shown that the film thickness can continue to increase even after 1 h immersion. However, for these reasons of concentration and dilution, the gelled film is found to be stronger and more coherent nearer the former, with the latex film towards the outer layer being more of a viscous liquid than a coherent gel, at least before the drying stages.

In some cases, this sequence of coagulant dip then latex dip can be reversed to latex dip then coagulant dip, and dipped rubber articles are made by first dipping the uncoated former into the latex and withdrawing (see straight dipping above), partially drying this first dip and then dipping into the coagulant bath. This has the effect of firmly gelling the initial dip, and also providing a layer of coagulant to build product thickness through subsequent latex dips. This sequence of alternating coagulant and latex dips is the only practical way to manufacture very thick articles such as electricians' gloves, which can be up to 3 mm thick or even more. A major potential problem with this multiple dipping process is the risk of delamination of the layers. To avoid this problem, the coagulant must be carefully formulated and each dip must be dry (but not too dry) before the subsequent dip is applied. No powder will be used in the coagulant, which will often be alcohol based, so as to ensure good wetting without the use of surfactants which could interfere with the interlaminar bonding.

The latex formulation for coagulant dipping will often be similar to that used for straight dipping, but as the latex viscosity is less important in achieving the required film thickness, the latex used for coagulant dipping can often be run at a lower viscosity than that for straight dipping. This lower viscosity will make the latex more stable to mechanical agitation and less likely to entrain air bubbles. This, in turn, will allow faster former entry speeds, although care will still need to be taken if the former has significant undercuts or pronounced roughening. Typically, the total solids content of the latex in coagulant dipping will be around 40 to 45% (maybe even lower) whilst that in straight dipping can be over 50%. However, this lower solids content means that shrinkage can be a problem and also increases the risk of splitting if the rubber film has a low wet-gel strength, as can be the case with some NBR latices.

7.3.5 Latex drying

Immediately after dipping, the former – by now often moved into a horizontal position or even well above the horizontal – is transferred into an oven to dry the latex film. It is usual to rotate the former during drying to prevent thickness variations in the product caused by the latex slumping during the initial stages of drying. Several heat sources can be employed – direct gas heating, radiant heating or hot air. Temperature control can be achieved with thermocouples mounted in the oven (the nearer to the product the better) or by infrared monitoring of the film surface. Note that if direct gas heating is used there is the possibility that nitrous oxides may react with any residual secondary amines from the accelerator system to give *N*-nitrosamines. Where this form of heating is employed it is recommended that regular testing for *N*-nitrosamines is carried out to monitor the situation, especially if the dipped products are medical devices or may be used in food handling.

The extent of drying is important for subsequent operations such as beading or further latex or coagulant dipping. If the film is too dry there will not be sufficient tack for successful beading, and overdrying of the film is the usual reason for delamination of multilayer dipped products. Any subsequent latex dip may often utilise residual coagulant from the preceding dip, so this coagulant, together with the rubber film underneath it, must be in exactly the right condition. Thicker products require careful drying. If a skin of rubber is formed over the surface whilst the latex underneath remains wet, various problems can ensue. The underlayer can remain wet, and hence weak, and be damaged by subsequent operations. If the temperature rises too high, the water in the still-wet layer can boil and give rise to bubbles ('overdrying bubbles') in the film. It is always better to make the drying progressive and/or moderate the latex pick-up where possible.

In those cases where shrinkage is a problem, for example, where the rubber has a low wet-gel strength, or the former is patterned and double imaging of the pattern occurs, then the drying will have to be gentle and well controlled to help reduce the problem.

7.3.6 Subsequent dips

Although some products are made in a single dip using coagulant dipping, in many cases more than one dip will be used, e.g., to build thickness or to provide a specialised coating. These dips can be latex or, in the case of some articles such as powder-free surgical gloves or catheters, a special polymer coating. There will usually be residual coagulant on the surface of the article after the first dip, and this is frequently used to gel a subsequent dip. However, as mentioned above, this residual coagulant will not be as concentrated nor as effective as that for the first dip, and to compensate for this the dwell time may be longer, the latex formulation may be made less stable (always risky) or the viscosity of the latex can be increased using hydrocolloid thickeners, such as carboxymethyl cellulose salts or polyacrylates. If the main objective of subsequent dips is to increase thickness, for example, in the manufacture of certain heavy-duty industrial gloves, further coagulant dips will be used before any latex dip, as this is the most effective way.

7.3.7 Flocking

Many houseware gloves and some industrial gloves are given a flocked coating to create a more comfortable inner lining and to make the gloves easier to put on and take off. Note that as gloves are almost always inverted during stripping, the outer layer will become the inner layer, so the flocked coating will be on the outside as dipped. In this case, the second dip will often function solely as an adhesive layer for the flock. As such it is not expected to contribute to the overall strength of the glove, and will be thin and often heavily loaded with filler to reduce the cost. This latex dip must have the correct balance of stability at the time of flocking – too wet and the flock will sink in and not give the soft coating. The flock can also absorb the latex which will give a very harsh feel to the surface. On the other hand, if the latex dip is too dry, the flock will not adhere and will be removed during subsequent processing operations. These requirements impose further constraints on the coagulant and latex formulations, and on the drying time and temperature. The material used for the flock varies, but will typically be short fibres of cotton or polyester, or a mixture of the two. The application technique can be simply blowing the material at the formers or, more expensively, electrostatic flocking can be used. In this process an electrostatic charge is applied to the flock fibres. The charged fibres will be attracted to the nearest earthed surface, and if the coated formers are in the way, they will become covered in the flock. Electrostatic flocking also tends to orientate the fibres so that they are vertical to the latex dip, enhancing the tactile sensation and effectiveness of the lining.

7.3.8 Beading

As with products formed by straight dipping, many dipped latex products will incorporate a bead at the open end to strengthen the edge, at what is generally the thinnest part of the product. Beading can take place either after the first dip, after any second dip or both. What is important is that the film must be at the correct stage of dryness to give a good bead. If the film is too dry it will tend to adhere to the former and may not be tacky enough to form the bead: if the film is too wet it will be weak and can tear. Excess moisture in the bead can also give rise to bubbles and 'blow' the bead during any subsequent heating. For small-scale production the bead can be formed by hand, but for most industrial processes it will be done automatically. The formers are rotated past rotating rollers and/or brushes that are at a slight angle to the main axis of the former. Other beading devices, such as rotating flexible rubber flaps may also be used. Several rollers or brushes will usually be used to give the correct bead dimensions.

Not all types of latex can be beaded. Some of the NBR latices, especially those latices with a higher ACN content, do not have sufficient surface tack to form a good bead.

As with straight dipped products, the bead (where present) is an important part of the product design, and bead defects must be minimised to ensure that a good yield of product is achieved off the production line. This in turn demands that strict attention is paid to the dryness and temperature of the rubber film at beading, and the design and accurate set-up of the beading brushes and rollers.

7.3.9 Leach

Leaching a wet-gel film, i.e., one that has not completely dried, is more efficient than leaching a dry film, as the water is able to penetrate the film more easily. Due to the need to remove as much residual coagulant and coagulant reaction products as possible, wet-gel leaching will be used in coagulant dipping whenever possible. The leaching stage will therefore be situated directly after the beading and any flocking/ drying operations, and before the vulcanising ovens. The leaching will be carried out by water, usually at an elevated temperature. It is common to have some discarding and replenishment of the leach water to prevent too much build-up of extracted materials in the water.

As described in the section on straight dipping, the rate and extent of leaching is dependent on the amount of prevulcanisation and leach temperature, as well as the type of rubber material. As with straight dipped products, an effective leach can improve physical properties and clarity of the finished product, but note that if any water-soluble accelerators, such as sodium or ammonium salts are used in the formulation, then the efficiency of any subsequent vulcanisation steps may be reduced.

7.3.10 Vulcanising

Although many latices used in coagulant dipping are prevulcanised to some extent, it is common practice to carry out a further heat treatment to consolidate the latex film and continue any crosslinking. The temperature and time of this 'vulcanisation' will depend on the product and materials, but will typically be up to 20 min at up to 120°C or maybe even higher with CR products. However, higher temperatures are not often employed because of the risk of thermal damage to the thin film and the risk of boiling residual water in the bead giving rise to blisters and a 'blown' bead. The risk of damage to the latex film caused by residual water can be avoided, or at least minimised, by gradually increasing the temperature through a sequence of ovens. Nevertheless, as some formers, especially the larger glove formers, have a large thermal capacity, high oven temperatures are required to attain the required film temperature. Some experiments have been carried out using high-humidity vulcanisation, where the initial stages of the drying ovens have the humidity artificially raised. This allows higher temperatures to be used to promote the crosslinking reaction (which proceeds more easily in the presence of water) without risking overdrying and/or blowing of the film. However, maintaining a high-humidity environment on a production plant, especially a chain plant, can pose considerable engineering and control challenges.

7.3.11 Prestrip soak

Passage through the vulcanising ovens will tighten up the latex film, as water is removed and the rubber shrinks and clings more tightly to the former. For that reason it is common practice to have a short soak tank after the exit from the vulcanising ovens and before the stripping station to soften the latex film and help release it from the former. This soak tank is usually hot water, although a small amount of alkali may be added to help the water release the film more easily. The addition of surfactant is best avoided, as this can make the product slippery, complicate stripping and generate foam.

7.3.12 Stripping

Depending on the product, stripping can be carried out manually or automatically. Automatic stripping can use water jets, rotating flaps or brushes, or some other form of mechanical device. A jet of compressed air is sometimes used to inflate the product momentarily, and to help release it from the former. One point to note is that the article will be inverted by some forms of stripping, so that the surface next to the former becomes the outside of the article. This is important if the article has some form of asymmetry, such as the left and right hands of a glove, where the former carries a pattern that is required to be on the outside of a glove, or where the inner and outer surfaces are different, for example, if a special coating or flock has been applied to one surface. Other types of stripping, such as water jets can strip the product off without inversion, or invert some and not others. The type of stripping mechanism and how it is set up is therefore very dependent on the product being dipped.

7.3.13 Former clean

Because coagulant dipping frequently uses higher temperatures and more textured former surfaces than straight dipping, especially if dipping gloves, soiling of the formers can be greater, and a more intensive cleaning regime may be needed. Typically, routine cleaning will be a hot water spray or tank after each cycle and there will also be the capability of a more intensive clean at intervals using an acid or alkali soak, often in combination with ultrasonic agitation. Weaker acids such as citric acid are used, although more powerful oxidising acids such as chromic acid may be necessary if the formers become heavily soiled. Sometimes a sequestering agent such as a salt of ethylenediaminetetraacetic acid will be employed to help remove contamination.

The intensity, frequency and type of former cleaning will depend upon the product, production and former type. As has been noted, clean formers are essential to minimise wetting problems and maintain quality, so the importance of proper former cleaning should be recognised.

7.4 Heat-sensitised dipping

Many latex formulations are temperature sensitive to some extent and become less stable as the temperature is raised. This is the result of increased energy of the individual latex particles, and the subsequent greater ease with which they can surmount the energy barrier that prevents coalescence. As discussed in Chapter 6, the zinc ammine reaction can also be brought into play to give a degree of heat sensitivity to a compounded latex. With care, this temperature sensitivity can be controlled and used in heat-sensitised dipping. It is possible to build up a thick deposit on the former with heat-sensitised dipping, and as such it will typically be used to manufacture thicker articles such as baby bottle teats, baby soothers, some industrial gloves, bladders and meteorological balloons, and footwear. It is almost always used with NR latex formulations. In general, the colloid stability of NR latex will be reduced by lowering the pH to around 8.0 to 8.5, usually increasing the concentration of zinc oxide, adding a destabilising agent such as ammonium acetate or a surfactant with a reverse temperature/solubility behaviour, i.e., one that becomes less soluble as the temperature is increased. Frequently, the optimum solution will employ some or all of these, and several possible formulations are described by Gorton and Pendle in [7]. Obviously, the ideal situation is one where the latex remains stable until the temperature threshold is exceeded, when it will rapidly coalesce to form a firm gel. However, this ideal situation is not met in practice, although some formulations come close enough to it to be eminently practical.

In general, the dipping procedures for heat-sensitised dipping will be similar to coagulant dipping, although it is more common to use a batch plant for this purpose. The formers will often be stainless steel, which has a good heat capacity. A high heat capacity will dictate how long heat can be delivered to the latex, which in turn will influence the thickness and uniformity of the gelled film that can be obtained. The formers will be heated to a carefully controlled temperature, typically 60 to 80 °C. Temperature control is important, and a tolerance of no more than ±1 °C has been proposed [2]. The temperature of the latex in the dipping tank will also be carefully controlled. Another important factor is pH, as stated above, and this will usually be reduced to the required level by the careful addition of formaldehyde. The formers will then be immersed in the latex and held there to allow the latex in the vicinity of the former to become destabilised, giving a gelled rubber film clinging to the surface of the former. As would be expected, leaving the formers longer in the latex will give a heavier pick-up and a thicker rubber film. Typical dwell times are 15 to 60 seconds, and articles up to 5 mm thick, or even thicker, can be produced in this way. The subsequent operations (leaching, vulcanisation, stripping, finishing and so on) will follow similar procedures to coagulant dipped products, although the extra thickness may mean that leaching times are longer and stripping can become more difficult.

More detailed information on heat-sensitised dipping is given by Gazeley, Gorton and Pendle in reference [6] and by Gorton and Pendle in [7].

7.5 Electrodeposition

The individual latex particles in an anionic latex – and virtually all latices used for dipping are anionic – carry a negative electrical charge. If an electrical field is applied to such a latex, the latex particles will be attracted to the anode. Once at the anode they will lose their electrical charge – an important component of their colloidal stability – and thus will tend to coalesce and form a deposit on the anode. As this deposit is still swollen with water (in an aqueous latex) the electrical pathway is still open, and latex particles will continue to be attracted to the anode, building up the thickness of the deposit. Experiments with this process have shown that even complex former shapes, with excessive undercuts which would be very difficult to coat using more established dipping techniques, can be coated using electrodeposition. However, electrodeposition using an aqueous latex has a major drawback: the current will also cause hydrolysis of the water, with evolution of oxygen at the anode, giving a porous rubber film. There can also be the risk of oxidation of the rubber. This can be prevented by reducing the electrical potential to the point where electrolysis of the dispersion medium does not occur, but under these conditions the rate of electrodeposition is so low as to be impractical. For these reasons the formation of rubber articles from a latex by electrodeposition is only of theoretical interest rather than practical. An expanded treatment of the electrodeposition of latices is given by Blackley [8].

7.6 Spray coating

From time to time interest is raised in spraying latices onto the former, as an alternative to dipping. Theoretically, such a manufacturing plant could be considerably smaller than a dipping line and more versatile. When correctly formulated, rubber latices can be sprayed, and this technique has been used for the small-scale preparation of rubber moulds and so on. However, there are considerable technical issues in spraying latices, not least the stability problems that will occur because of the shear forces to which the latex will be exposed. At the time of writing, the author is not aware of any significant commercial use of this technique. Dipping latices is a very simple and well-understood process, and there appears to be little incentive to complicate the process unless the rewards are worthwhile.

References

1. R. Burley and R.P.S. Jolly, *Chemical Engineering Science*, 1984, **39**, 1357.
2. D.M. Bratby in *Polymer Latices and Their Applications*, Ed., K.O. Calvert, Applied Science Publishers Ltd., London, UK, 1982.
3. K.F. Gazeley, *NR Technology*, 1985, **16**, 61.
4. A.D.T. Gorton, *NR Technology*, 1979, **10**, 9.
5. R.W.T. Merrill in *Prevulcanization of Natural Rubber Latex*, London School of Polymer Technology, The Polytechnic of North London, UK, 1980 [MPhil. Thesis].
6. K.F. Gazeley, A.D.T. Gorton and T.D. Pendle in *Natural Rubber Science and Technology*, Ed. A.D. Roberts, Oxford University Press, Oxford, UK, 1988.
7. A.D.T. Gorton and T.D. Pendle, *NR Technology*, 1980, 11, Part 1, 1.
8. D.C. Blackley in *Polymer Latices Science and Technology*, Chapman and Hall, London, UK, 1997.

8 Wetting

8.1 Wetting of a solid surface by a liquid

The ability of the latex or coagulant solution to wet the former is fundamental to latex dipping. If a uniform coating of the latex or coagulant over the former cannot be achieved, then a good dipped product cannot be made. For this reason, controlling the wetting ability of the latex compound is one of the key aspects in developing a latex formulation, and a superficial knowledge of wetting can help the latex chemist.

Wetting is defined as the ability of a liquid to maintain intermolecular contact with a surface. It is the result of the balance of 'adhesive' forces and 'cohesive' forces in play within the system. The cohesive forces in a liquid are responsible for the phenomenon known as surface tension or surface free energy. In a drop of liquid these cohesive forces seek to attain a minimum value for the free energy at the liquid–air interface, and thus the drop will adopt a configuration with the minimum surface area (a sphere), hence the reason why drops of liquid are spherical. If the liquid is placed in contact with a solid surface, another set of forces comes into play: the adhesive forces between the liquid and the solid. These adhesive forces will try to maximise the contact between the solid surface and the liquid, and as such will try to spread the liquid out over the surface, whilst the cohesive forces within the liquid will try to maintain the spherical configuration. The balance between these forces will dictate how easily the liquid spreads (or not) over the surface; that is, how easily the surface is wetted by the liquid. The balance between these two forces can be best envisaged and measured by the 'contact angle' (θ) of a drop of liquid on the surface. The contact angle is the angle made by the liquid–vapour interface of the drop with the surface of the solid. This angle, which is by convention measured through the liquid, gives a way to quantify wetting and wettability. Figures 8.1 and 8.2 show two different extremes.

In Figure 8.1 the drop of liquid is not wetting the surface. This situation will be familiar to anyone who has seen water on a greasy plate or raindrops on a leaf. The surface free energy of the liquid is greater than that of the solid surface and the cohesive forces within the liquid are greater than the adhesive forces between the liquid and the solid surface. Thus, the liquid is trying to minimise contact with the solid and the contact angle is large. The opposite situation is seen in Figure 8.2 where the liquid is spreading out over the surface and contact angle is small. Here, the adhesive forces between the solid surface and the liquid are greater than the cohesive forces trying to hold the liquid as a drop, and the liquid will spread out over the surface. In this situation the surface free energy of the surface is higher than that of the liquid. Another way of looking at it is that the change in energy of the system

https://doi.org/10.1515/9783110638097-008

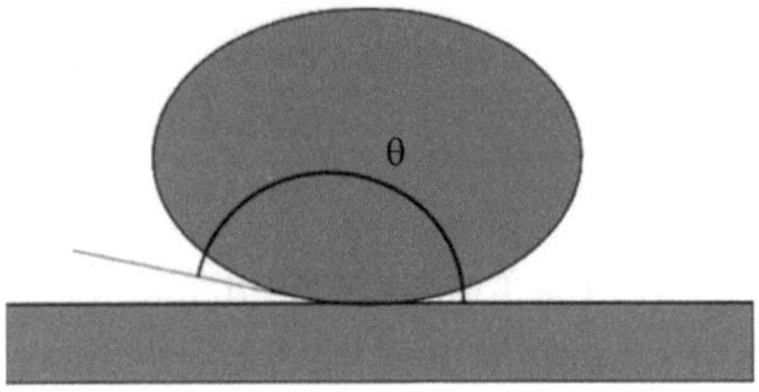

Figure 8.1: A diagrammatic representation of poor wetting – the contact angle (θ) is large.

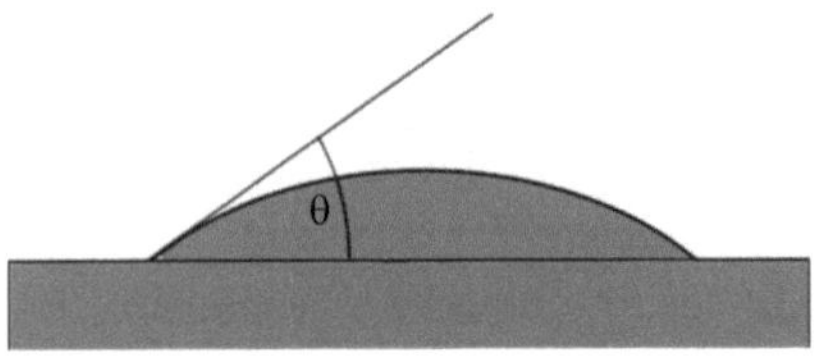

Figure 8.2: A diagrammatic representation of good wetting – the contact angle (θ) is small.

must be favourable, that is, the total energy must be reduced. In the case of natural rubber (NR) latex, the surface free energy will typically be in the range of 30 to 40 mN/m, depending on the formulation and the concentration of surface active agents. The surface free energy of, for example, glass, is typically around 70 mN/m. Thus, coating the surface of the glass with latex is energetically favoured and glass is easily wetted by latex. On the other hand, some fluorinated plastics have surface free energies of around 15 mN/m. Covering the surface of these materials with latex will not be energetically favourable, and our NR latex – or in fact any latex – will not wet the surface of such materials. It is important to realise that any contamination on the surface of the former will change the surface energy and have a direct effect on wetting. Approximate values of surface free energy for some solids and liquids are listed in Tables 8.1 and 8.2, respectively.

Some points of interest to note from these figures:

- Many of the plastic materials have a surface energy similar to or lower than compounded latex, and so would not be wet easily without modifications to the latex formulation or the plastic surface.
- Generally, glass and ceramic materials have a high surface energy and will be easily wetted by the compounded latices.
- A water-based coagulant solution will generally need the addition of a surfactant to lower the surface energy in order to wet the formers well.
- A clean NR surface has a low surface energy and will therefore be difficult to wet. The various surface-active materials in compounded NR latex mean that the actual surface energy of a dipped rubber article will be higher, but even so the second dip of, for example, a condom, will not wet as easily as a first dip onto a glass former.

Table 8.1: Some approximate values of surface free energy for solids.

Solids	Surface free energy (mN/m)
Acrylonitrile-butadiene- styrene terpolymer	42
Aluminium	Approximately 35
Ceramics	Approximately 40–75, depending on the composition
Glass	Approximately 43–112, depending on the composition
NR	18
Polycarbonate	Approximately 29–39
CR	44
Polyethylene	Approximately 27–36
Polyethylene terephthalate	Approximately 37–47
Polymethyl methacrylate	Approximately 39–46
Polypropylene	Approximately 28–34
Polystyrene	Approximately 29–43
Polytetrafluoroethylene	Approximately 16–22
Polyvinyl chloride	Approximately 27–44
Steel	Approximately 40–50

CR: Polychloroprene
Reproduced with permission from KRÜSS GmbH. ©KRÜSS GmbH [1]

Table 8.2: Some approximate figures for the surface free energy of liquids.

Liquids	Surface free energy in mN/m
Ethanol	Approximately 24
Polydimethylsiloxane	22
NR latex	Approximately 30–40, depending on the formulation
Acrylonitrile-butadiene latex	Approximately 28–40, depending on the formulation
CR latex	Approximately 35–42, depending on the formulation
Synthetic *cis*-1,4- polyisoprene latex	Approximately 35
Water	Approximately 72

Reproduced with permission from KRÜSS GmbH. ©KRÜSS GmbH [1]

As well as the ease of wetting, the speed of wetting is important. In most production situations, the speed of wetting is fast enough not to be a problem, but in some cases (e.g., in the production of examination gloves) the plant speeds are quite high, and the speed of wetting will need to be taken into consideration. The main potential problem is that air will be entrained into the latex if the entry speed is faster than the speed with which the latex wets the former, and air bubbles in the latex are a potential source of defects and best avoided. This speed of wetting of formers can be calculated, for example, by work done on the coating of magnetic tape media [2]. The following formula, Equation 8.1, was derived by Burley and Jolley for the critical speed of entry:

$$\text{Entry velocity}(\text{cm/sec}) = 70.5\,[\mu\,(g/\rho\sigma)^{0.5}]^{-0.77} \tag{8.1}$$

where 'μ' is the latex viscosity, 'g' is the acceleration due to gravity, 'ρ' is the latex density and 'σ' is the latex surface tension.

In a typical NR latex system, we may assume that $\mu = 25$ cps, $g = 9.81$ m/s^2, $\rho = 0.94$ and $\sigma = 35$ mN/m. Inserting these values into the equation gives a critical entry speed of 9.4 cm/s to avoid air entrainment. If the former enters the latex faster than this speed there is the likelihood that air will be pulled into the latex.

It is always a good idea to keep a close eye on how a latex dipping plant is behaving. For example, observation of the meniscus at the interface between the latex and the former as the former enters into the latex can yield valuable information about the cleanliness of the former and the evenness of the wetting. If the surface of the latex dips down erratically as the former enters the latex, and seems reluctant to wet the former surface, then it is likely that the surface of the former is not clean enough. If the latex seems to 'climb up' the former as it enters the latex, then it is likely that the surface of the former is damp. A clean, dry former will wet smoothly and evenly.

8.2 Measurement of wetting

There are several ways to measure and quantify wetting and surface free energies. For liquids, the surface free energy can be determined by measuring the force required to pull a metal ring out of the liquid – the Du Noüy method, described in International Organization for Standardization (ISO) 1409:2006. Alternatively, the actual contact angle made by a drop of liquid on a surface can be measured optically using a goniometer. The contact angle is represented by the Greek letter 'θ', and conventionally is expressed as the angle measured through the liquid. It is usual to measure the contact angle after the drop size has been increased (the 'advancing angle'), and also after the drop size has been reduced by sucking a little liquid up into a syringe (the 'receding angle'). Once the contact angle is known, the interfacial energy can be calculated using, for example, Young's equation [3], Equation 8.2, which gives a relationship between surface free energy of the solid, 'σ_S', the interfacial tension between the liquid and the solid, 'σ_{SL}', the surface tension of the liquid, 'σ_L', and the contact angle (θ):

$$\sigma_S = \sigma_{SL} + \sigma_L.\cos\theta \tag{8.2}$$

Young's equation assumes that the system is in thermodynamic equilibrium, with a pure liquid and a smooth rigid surface. As these limits are not often met in practice, there are a number of other models that attempt to calculate the surface free energy from the contact angle using additional terms to account for the interaction between the phases. The interested reader is referred to the wealth of science in the literature on the subject.

There are other ways to measure these properties, e.g., the surface free energy of a liquid can be measured from the actual shape of a drop of liquid. These techniques use sophisticated analytical instruments, and are likely to be found more in research laboratories and universities than in a latex dipping factory. For the production of dipped latex articles, the Du Noüy equipment or a simple goniometer will be perfectly adequate to help the manufacturer understand wetting and wetting-related problems.

References

1. www.kruss.de/services/education-theory/substance-data.
2. R. Burley and R.P.S. Jolly, *Chemical Engineering Science*, 1984, **39**, 1357.
3. T. Young, *Philosophical Transactions of the Royal Society of London*, 1805, **95**, 65.

9 Latex dipping tanks

The general outline of latex dipping lines, and the different elements that comprise them, have been given in Chapter 7. This chapter will discuss in more detail the dipping tanks used on a dipping line and the means by which the formers are dipped into the tanks. This is a critical part of the manufacturing process, and one which, if correctly designed and utilised, will be of considerable help in producing good-quality product.

As discussed in Chapter 7 there are basically two different designs of dipping lines in use: chain plants and batch plants.

9.1 Chain plants – general

Chain plants consist of an endless chain driven by a motor (sometimes two motors on larger plants, one at each end of the line). The formers are mounted onto the chain and are carried by the chain around each of the stages in the dipping operation in turn. Chain plants can be double-sided, with formers mounted on either side of the chain, or single-sided. If the plant is double-sided, it offers the possibility of manufacturing different products on each side, although of course the line speed will have to be the same for each product. Double-sided plants may use the same dip tank, or each side may have its own dip tank.

The position of the drive motor is important. It should be located so that the chain is pulled through the more tortuous path (typically the lower level, with the entry and exit cams into the various dip tanks) and pushed through the straighter zones – often the drying ovens. This means that there is less tendency for the chain to buckle, which can lead to formers touching each other. Some large glove plants, however, can have extended drying zones where the chain can snake through the ovens. In this case, the drive motor will need to be located to minimise the possibility of any chain buckling. In extreme cases, two motors may be needed.

The former spacing on the chain must be sufficient so that the formers do not touch each other, especially where the chain goes around sharp curves, and also to allow for those formers where there may be some slight misalignment in the mounting.

It is especially important to align formers correctly where the formers are long, e.g., with some industrial gloves.

9.1.1 Entry/exit mechanisms on a chain plant

In order to direct the formers, whether into the latex, out of the latex, into the drying ovens and so on, the formers will run along a track (often called the 'cam').

https://doi.org/10.1515/9783110638097-009

Where the formers are required to be lowered into a tank (e.g., the coagulant dip, latex dip or leach tank) the chain track will be directed downwards, and similarly upwards to raise the formers out of the tank. This will be best accomplished by attaching the formers to the chain through a holder that can rotate and pivot (a 'spinner'), with the holder being the contact point between the former and the cam. An example of this on condom formers can be seen in Figure 9.1, showing condom formers emerging from the leach tank.

Figure 9.1: Condom formers on a chain plant emerging from the leach tank. Reproduced with permission from the Thai Nippon Rubber Industry Public Company Limited. ©Thai Nippon Rubber Industry Public Company Limited.

The 'spinner', the collar holding the former that is free to rotate, can clearly be seen, as can the cam along which the spinner is rolling. Note also the ring on the spinner, which provides the contact onto the cam. These rings can be made to be replaceable, when they wear, but note that the latex dip tanks should be protected from any debris resulting from wear on the spinner contact rings. As can be seen from close inspection of Figure 9.1, the formers towards the right of the picture are at a slight angle to the cam, and thus there will be a 'scrubbing' action to some extent on the contact rings as the formers are carried through. This will occur at every point where the formers change attitude, and it is at these points that there will be the greatest risk of debris being worn off the spinner rings and precautions should be taken to minimise the risk of contamination. For this reason, the dip tank may be covered at each end to catch any debris, although it can be difficult to ensure complete protection without running the risk of the formers touching any covers as they enter and exit the latex.

Figure 9.2: An example of the way in which condom formers can be attached through the spinner to the chain. Reproduced with permission from the Thai Nippon Rubber Industry Public Company Limited. ©Thai Nippon Rubber Industry Public Company Limited.

One way of attaching the spinner to the chain is shown in Figure 9.2. In this design, the spindle of the spinner is locked into the holder by a grub screw. The holder carries a hinge that allows the former to pivot (in this example towards the camera) against a spring that applies a counter force to help maintain the spinner in contact with the cam. This is just one example of the way formers can be mounted on the chain, but other designs, both for condoms and gloves, will follow similar principles, if not the actual design.

It is important that the chain movement is smooth, especially during immersion and withdrawal of the formers. Any juddering or other rapid changes in speed can help drag air into the latex as the former enters, and will cause variations in pick-up on exit from the latex. It can also increase wear on the cam and other parts of the machinery. If the exit profile of the cam is curved, then the latex pick-up in straight dipping can be modified to help give a more even wall thickness, as discussed in Section 7.2.2.

There will also be some way of running the plant with the formers held out of the dip tank ('dipped out'), to allow access to the dip tanks. This will usually take the form of an extra rail over the dip tanks, along which the formers can run without being immersed in the latex. There will be some form of gate to switch the path of the formers, and it is important that operation of this gate is safe. With a double- sided plant care must be taken when dipping only one side of the line out. In this situation there will be a greater load on the side of the plant that is dipped out, as the buoyancy experienced by the formers when immersed in the latex will be absent. Where the formers are heavy, for example, on an industrial glove plant, these forces may lead to distortion of the frame of the dipping line.

In general, chain plants have the dipping tanks located at ground level, but the lighter drying and vulcanising ovens are often located above.

9.2 Batch dipping plants

Batch dipping plants provide a complete contrast to chain dipping plants. In a batch plant, the formers are mounted on boards or jigs, which can contain anything from one former to several dozen or more. The formers are immersed in the latex, either by the formers being lowered into the latex dip tank, or the tank being raised to the formers. Usually a hydraulic or mechanical mechanism is used for these movements, and it is easier to control and vary the speed and extent of motion during dipping on a batch plant than on a chain plant. After dipping, the formers will be transported to the next stage in the manufacturing process.

Batch plants have both advantages and disadvantages over chain plants. They are usually smaller and the dip tanks can be shorter, so will not be subject to all of the inherent problems of agitation in a long thin tank. Multiple dips can be achieved with only a dipping tank and one oven. They are more versatile and can accommodate changes in the manufacturing process, such as entry/exit and dwell times, more easily. The disadvantages are that operations such as beading and stripping can be more complicated, and the output is generally much lower. It will usually be found that high-volume manufacturing, such as condoms, examination gloves, surgical gloves and so on, will be produced on a continuous chain plant, whilst heavyweight industrial gloves, baby bottle teats and so on, will often be made on a batch plant.

A minimum-sized batch plant will be found in laboratories carrying out development work on latex dipping. These will typically carry only one or two formers, and are only required to raise and lower the former. They will usually be driven by an electrical motor or compressed air. They can be very simple, with the entry and exit speeds controlled by hand, or more sophisticated with full computer control. Drying and vulcanising will usually be carried out in an adjacent laboratory oven. Condoms can be dipped in a measuring cylinder and gloves in a larger dipping pot, e.g., constructed from a polyvinyl chloride (PVC) drainpipe.

9.3 Latex dipping tank construction

A major consideration in the design and layout of a dipping line are the dipping tanks, especially the latex tanks. Dipping tanks on a latex dipping line can vary widely in size and design, depending on whether they are on a continuous chain plant or a batch plant; and whether the product is made using coagulant dipping, straight dipping or heat-sensitised dipping. In the majority of cases the tanks will be constructed from stainless steel, although hard plastics such as reinforced polymer resin or PVC are also used in some cases.

One important factor to consider in the design of a dipping tank and associated equipment is to avoid any use of copper. Whilst copper is unlikely to be used as a main construction material, it may be present in pipework or in phosphor–bronze bearings in agitators or pumps. Even tiny amounts of copper can cause staining in the presence of dithiocarbamate accelerators, and copper can readily promote the oxidative degradation of natural rubber. The same applies to manganese and cobalt, although these are less likely to be encountered in these situations.

9.4 Latex dipping tank design

9.4.1 Dipping tanks for straight dipping

Straight dipping dictates that the latex pick-up, and hence product thickness, is controlled by, amongst other things, withdrawal speed. On a continuous chain plant this means that the dipping tank must be long (frequently several metres long) in order to give the correct withdrawal speed with an acceptable line speed. This in turn brings in various practical problems, such as agitation and circulation of the latex. Agitation is essential in a latex dip tank for several reasons, including the homogenisation of temperature and composition, and preventing the latex skinning over. The ideal tank shape for effective agitation is circular: a long thin tank requires much more vigorous agitation in order to get effective circulation of the liquid. Latices are shear-sensitive materials, and too much agitation will quickly lead to the formation of coagulum and the inclusion of air bubbles. For these reasons, the effective agitation of a tank for straight dipping poses considerable problems, which do not yet appear to have been fully overcome. A typical design for a dipping tank used in straight dipping is shown in Figure 9.3.

This design carries a propeller-style agitator at one end (often, but not always, the exit end) circulating the latex through a false bottom or another return system such as external piping. An Archimedian screw can be used as an agitator instead of the propeller and the agitator is often enclosed in a shroud to enhance efficiency. The clearance around the agitator is important: too little clearance can give rise to coagulum because of the shear forces that will be generated, whilst too much

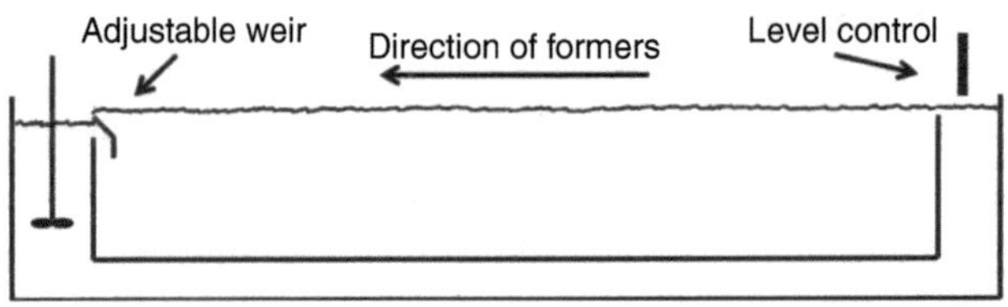

Figure 9.3: A sketch of a typical design of a latex tank for straight dipping.

clearance will give inefficient circulation. It is better if the speed of the agitation can be varied. Even so, the rate of circulation of the latex in such a tank can be very slow and the bulk of the circulation results from the movement of the formers through the latex. This situation, where the formers are moving faster than the latex creates turbulence, leads to variations in latex pick-up, as discussed in Section 7.2.2.

Another design of dipping tank which is sometimes used in straight dipping lines is the 'island' design (Figure 9.4), where the latex is circulated through a separate return channel, usually covered to reduce water loss through evaporation.

Figure 9.4: A sketch (plan view) of a typical island dipping tank for a straight dipping line.

In this design, agitation is provided solely by the movement of the formers through the latex. This type of dipper is most effective on dipping plants with a high line speed, and can offer a solution to some of the problems associated with dipping tank design.

The latex in the dip tank is therefore exposed to various destabilising influences, such as agitation, shear, heat from hot formers in the second dipper and evaporation of water from the surface. It is therefore desirable to keep the volume of the latex in the dip tank as low as practical, so that the latex is continually being replaced and the residence time of the latex in the tank is short. However, this brings about its own problems: the motion of the formers into, through and out of the tank can set up a standing wave within the tank, causing the surface of the latex to rise and fall in a regular pattern. The smaller the tank, the more likely this is to occur.

If the latex dipper is affected in this way, the standing wave can easily be seen by looking along the surface of the latex whilst the line is running. The presence of a standing wave in the dip tank will modify the effective withdrawal speed – that is the rate at which the former leaves the surface of the latex. Assuming the rate of vertical movement of the formers is constant, when the standing wave causes the latex surface to rise, the effective withdrawal rate is reduced, and as the latex surface falls the withdrawal speed is increased. As the latex pick-up in straight dipping is affected directly by the withdrawal speed of the former, the consequence of this standing wave is a fluctuating pick-up of the latex and therefore a sequence of thickness variations, which can often be seen as a pattern of bands down the dipped article as it dries. Although variations in thickness can lead to cracking of the latex film, in practice, the pattern of thickness variation caused by the standing waves does not often lead to problems, unless the product is very thin.

The formers entering the latex in the dip tank are frequently hot, either from the drying of a subsequent latex dip or drying after cleaning. This will tend to warm up the latex, reducing the viscosity and increasing the rate of evaporation of water from the surface. Increasing the latex temperature can also reduce latex stability. It is therefore common practice to control the temperature of the latex in the dip tanks. This can be achieved by jacketing the tank and passing cooling water through the jacket or by having a cooling coil located in the latex tank itself. In both cases good agitation is required, otherwise there will be temperature variations within the latex, giving localised warm and cold zones that will affect the viscosity. Owing to the adverse effect that an elevated temperature can have on latex, it is advisable that formers entering the latex are not too hot. It is not possible to specify any maximum temperature, but it is recommended that the temperature of formers entering a latex does not exceed 45 °C, and is preferably below 40 °C. This also ensures that the cooling fluid does not need to be too cold, hence temperature gradients between the latex and the tank wall are reduced.

Agitation in the dip tank has another important function, i.e., keeping the latex as homogeneous as possible. Not only is there water loss from the surface of the latex, and temperature variations resulting from the latex being heated by hot formers and cooled by a cooling coil or jacket, but many of the ingredients of a latex formulation have a specific gravity greater than that of the latex. The densities of some of the more common components of a latex formulation are listed in Table 9.1.

Table 9.1: The densities of some of the more common components of a latex formulation.

Material	Specific gravity
Sulfur	Approximately 2
ZnO	5.6
Antioxidants	Typically 1.05–1.2
Zinc diethyldithiocarbamate	1.46
Zinc dibutyldithiocarbamate	1.26
Zinc mercaptobenzothiazole	1.64

ZnO: Zinc oxide

Thus it can be expected that solid materials such as ZnO, sulfur and accelerators will sediment rapidly in latex unless there is sufficient agitation to maintain them in suspension. As ZnO and zinc-based accelerators can destabilise many latices, any localised high concentrations of these materials, such as will occur in and around any sediment, will lead to the formation of gel and coagulum. For this reason both effective agitation and regular cleaning of the dip tanks are important. Filters, such as a Nylon or stainless steel mesh are sometimes placed in the recirculation path of the latex, to trap any coagulum or skin and prevent it being picked up by the product. The effectiveness of these is doubtful, as the restriction they cause slows down the latex circulation, which in turn leads to the formation of skin and cream, which clog the filter and reduce the circulation even more.

'Bubble traps', small containers containing baffles (Figure 9.5) that hold back any bubbles on the surface of the latex – can also be fitted to the latex feed into the dipping tanks and will help if the production process has a tendency to generate bubbles.

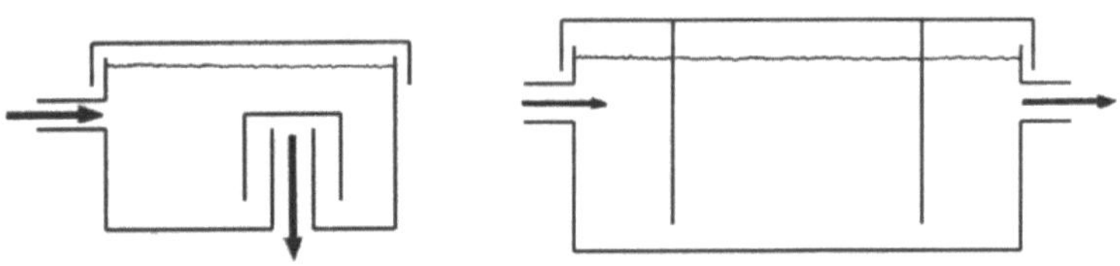

Figure 9.5: Two designs of a bubble trap – the direction of latex flow is indicated by the arrows.

Control of the level of the latex in a dipping tank is important, as variations in the latex level, leading to variations in the dipped length on the former, can give rise to beading problems. In many cases, the level is controlled manually with the correct level being indicated by a pointer or some other kind of marker. For straight dipping, the position at which the former leaves the latex surface can also be used to indicate the correct level of latex. In these cases, the latex feed will usually be regulated so that the latex continually trickles in at a rate approximately equal to the latex consumption, and the level will be checked and adjusted at regular intervals by the plant supervisor or other operative. It is also possible to fit automatic level controls onto a dip tank that open a valve in the latex feed when the level drops, and shuts it off again when the correct level has been reached. These will usually be optical devices, as devices based on a float can easily become fouled with a build-up of rubber. With automatic level controls, the sensor must be protected from the effect of any standing waves, otherwise the valve will be continually opening and shutting as the latex level falls and rises. This can be achieved by locating the sensor in a separate small well in the main tank. For the same reason – disturbance of the latex surface – the sensor is best located at the other end of the tank from the agitator. It is also important to ensure that the latex feed is turned off before the formers are 'dipped out' of the latex, otherwise latex will flood into the tank as the displacement of the formers is removed. In this situation, the latex can easily overflow when the formers are dipped back in again.

Level control is also assisted by having an adjustable weir over which the latex flows, which will smooth out any level variations.

As well as feeding latex in to the dipping tanks, it will also usually be necessary to trickle in a small amount of water (preferably ammoniated to approximately the same pH as the latex) to compensate for any water lost through evaporation, and to maintain the required total solids content of the latex. As with the latex feed, the formation of bubbles should be avoided.

Dipping tanks need to be cleaned at intervals and the design should facilitate this. Any cooling coils and false bottoms should be easily removed, and the agitator (if present) should be able to be easily removed and dismantled for cleaning. The inside of the tank should be smooth with a good surface finish to help cleaning. Wherever possible, edges should be rounded with no 'dead' spots where the latex is not circulated. There should be a drain plug, conveniently located, and it helps if the tank is on wheels so that it can be drawn away from the dipping line for cleaning or other maintenance. This, in turn, requires free access to the dip tank, at least on one side of the line. As latex dip tanks can contain several hundred litres of liquid, some form of mechanical assistance may also need to be accommodated. In addition, it can be useful to have jacking screws on the bottom of the tank, to stabilise the tank by raising the wheels off the ground and allowing the tank to be levelled. Jacking screws also allow the height of the tank to be adjusted to some extent, so that the immersion length on the formers can be controlled when necessary and to ensure that there is adequate clearance between the formers and the ends of the tank at the entry and exit, although in batch dipping lines, where the dipping tank is raised up to the formers, this will not be necessary. There should be provision to cover the surface when the tank is not in use, to help stop the latex skinning over. This can take the form of specially fitted covers, or a flexible plastic film can be laid directly onto the latex surface.

9.4.2 Dip tanks for coagulant dipping

The dip tanks for coagulant dipping can be made much smaller than for straight dipping, as the latex pick-up is controlled mainly by the coagulant, so that the withdrawal speeds can be faster. These smaller tanks are much easier to agitate efficiently than the much longer ones, hence the problems of sedimentation, air entrainment, coagulum, skinning and temperature variations are easier to overcome. The general principles outlined in the previous section concerning materials of construction, good agitation, temperature control, latex supply, ease of cleaning and so on, still apply. This type of dip tank can also be used on batch plants for both straight and coagulant dipping.

9.4.3 Dip tanks for heat-sensitised dipping

The most important consideration for heat-sensitised dipping is constancy of temperature, which has been stated should be within ±1 °C [1]. Hence, good agitation, effective jacketing of the tank and accurate temperature control are important, and most heat-sensitised dipping will be carried out on a batch plant, using the shorter, squarer dip tanks that this process can allow. The general principles described earlier regarding construction, latex supply and so on, will apply to these tanks as well.

Reference

1. D.M. Bratby in *Polymer Latices and their Applications*, Ed., K.O. Calvert, Applied Science Publishers Ltd., London, UK, 1982.

10 Formers for latex dipping

10.1 Former materials

The former, also called the mould or mold, is the template upon which the dipped article is made, and clean, undamaged formers are essential. Many materials have been and are used to make formers, including glass, ceramic materials such as porcelain, aluminium, stainless steel, wood, plastics and plaster of Paris. There are several requirements that the former material must meet in order to be dipped successfully. Firstly, it must be able to be formed into the required shape, and have a good level of physical strength and abrasion resistance. It must have a high surface energy relative to the liquid, so that wetting of the surface by the latex or coagulant solutions is easy, smooth and uniform. Typical surface energies of several solids and liquids are given in Chapter 8. Thermal resistance is important, as is resistance to alkaline solutions. Materials with a low coefficient of thermal expansion will be easier to mount securely onto the dipping machine, as the mounting will not be subject to as much expansion and contraction as the formers cycle through the hot and cool zones of the production cycle. Formers for continuous production must also be able to be cleaned repetitively without altering the surface characteristics.

Porcelain meets these requirements very well, and for this reason it is the preferred former material for many applications. Complex shapes can be created by slip-casting the ceramic slurry and formers can be made with a variety of surface finishes: glazed, roughened, patterned and so on. Porcelain is generally the material of choice for the formers in the production of gloves and balloons. Porcelain formers are usually hollow, but the wall thickness may not be consistent, for example, in glove formers the wall thickness is typically thicker in the fingertips than in the palm. Where present, this inconsistency may cause problems in, for example, heat-sensitised dipping, where the thicker areas of the former can retain the heat longer, and may give rise to uneven latex pick-up. Other ceramic materials, for example, those materials which have some type of molecular dipole allowing them to be heated by microwave or radiofrequency radiation, have been used experimentally; although as yet there is no widespread industrial application of this technology in latex dipping.

Glass is similarly widely used as a former material, although it is less easily formed into complex shapes than porcelain. It is the material of choice for condom formers. Whilst soda glass is cheaper and more easily formed than borosilicate glass, the borosilicate material has many advantages, including being stronger, having a higher resistance to alkali and a greater resistance to thermal shock owing to its very low coefficient of thermal expansion. It is also lighter than soda glass.

https://doi.org/10.1515/9783110638097-010

Aluminium is light and strong, but can be attacked by alkali, and as virtually all dipping latices are alkaline (some such as polychloroprene are highly alkaline), the risk of corrosion and pitting limits the use of aluminium.

Stainless steel is frequently encountered as a former material and has several useful properties. It is tough, can be made with a very high surface finish and is a better conductor of heat than porcelain or glass, making it more suitable for heat-sensitised dipping. However, stainless steel is very much heavier than glass or porcelain, although the strength of the steel means that the formers can be made thinner than those from other materials. Steel, and other metals, can also be scratched or dented giving flaws that will be reproduced in any article dipped from that former. Both glass and porcelain tend to break under a hard impact, and a broken former is easier to identify than a scratched or dented one!

Plastic formers have been used from time to time, and can be formed into complex shapes and are light. In addition, the emergence of three-dimensional printing technology means that formers can be produced from computerised designs in a matter of hours, which can be of great help in new product development. The wide variety of plastics available means that characteristics such as surface energy can be tailored to that required. Whilst modern plastics can be made very resistant to creep, their thermal resistance is still inferior to glass, porcelain and steel when subjected to the heat cycling experienced on a high-volume production plant. They are also more easily damaged than these harder materials.

Wood and plaster of Paris formers tend only to be used for prototypes. They are very useful for development work in the laboratory, but are not suitable for continuous production. Both of these materials need to be finished with a suitable lacquer to seal the surface and give an appropriate surface energy.

Whilst most former materials will last for many years, eventually the surface will become damaged by the cleaning regime, occasional knocks and other factors. A replacement schedule is recommended, although the replacement interval will depend on the particular circumstances. It can help if the formers carry some identification showing the date of installation, either engraved on the top of the former, or in the case of glass formers, a label can be placed inside the former.

10.2 Former design

Many former materials can be made with different surface finishes. Formers for industrial gloves, for example, can carry a grip pattern over the fingers and palm; and surgical glove formers are often lightly spray roughened or sandblasted, so that gloves made from these formers will have an enhanced grip capability. For some other applications, a smooth surface will be preferred in order to minimise the possibility of the former introducing flaws into the product. A roughened former surface will pick-up more coagulant or latex than a smooth surface, and will therefore give a thicker product. Some condom formers are deliberately made incorporating ribs, dots or more extreme patterns. However, note that former patterns which produce a thick zone of latex next to a thin area can be prone to cracking of the film as the latex dries (Chapter 6) so sudden changes of thickness are best avoided. The direction of any ribs must also, as much as possible, be aligned so that air can easily escape when the former is dipped. Formers can also carry some identification engraved on the surface, e.g., a number that will be reproduced on the dipped article. This can help locate any damaged formers giving rise to flaws on the finished product. Similarly, the manufacturer's logo or other symbols can be used. Note however, that the depth of any such pattern should be shallow to avoid problems arising from sudden changes of thickness, as described above.

Any former that has an undercut, which could trap air, will be more difficult to dip successfully. For this reason, the design of glove formers in particular must be carried out with care, also taking into account the angle of entry of the former into the liquid on the dipping line. The shape and curvature of the crotch between the fingers, and fingers and thumb, for example, is critical. Another element in the design of glove formers is the spacing between the fingers, which will affect the tendency of the former to create webs of coagulant or latex between the fingers during dipping. As each situation (former material, latex formulation, product design and so on) is different, no specific advice can be given here, other than to be aware of the constraints and factors when designing formers.

Any large changes in the former shape can also cause problems. As mentioned earlier, the pick-up, whether of latex or coagulant, is influenced by withdrawal speed. Curvature of the former (an extreme example is the end of a plain-ended condom former) will increase the effective withdrawal rate – the rate at which the former surface leaves the liquid, even though the linear withdrawal rate does not change (Figure 10.1). This results in a greater pick-up of coagulant or latex, leading to thickness variations and an increased risk of cracking. It is recommended that, wherever possible, the former design should try to grade changes in curvature to minimise this problem.

10.3 Former mounting

The formers need to be mounted onto the dipping chain or boards, and the basic requirements of these mountings are that they should be strong and reliable, but easy to replace the formers when necessary. For most dipping applications it will also be necessary for the formers to rotate and pivot away from the vertical to help distribute the latex evenly around the former, as described in Chapter 9. There are many different ways in which this can be accomplished – too many to describe in detail here. However, the fundamental points to observe when mounting formers can be listed as follows:

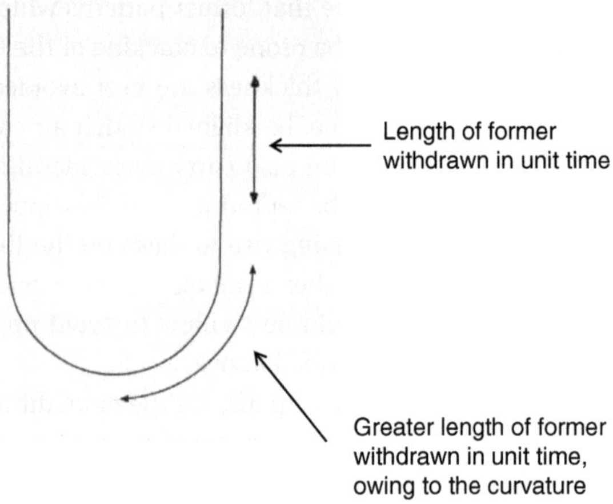

Figure 10.1: Sketch showing the increase in withdrawal rate that can result from curvature of the former.

- The former will be subjected to temperature cycling, sometimes over a temperature range of 100 °C or more. Rubber plugs, which are often used in compression fittings, will gradually harden under these conditions and may become very difficult to remove with time. It is therefore advisable to have a regular replacement schedule for such items.
- The former mounting can also be subjected to a range of chemical treatments, acids, alkalis, water, surfactants and so on, and therefore must be able to withstand corrosion.
- Replacement of a damaged former on a chain plant will often be carried out whilst the dipping line is still running, and therefore must be simple enough to be safely carried out on a moving chain.
- Replacement of the entire set of formers (e.g., if the production plan calls for a different product type) must be able to be accomplished with minimum downtime of the dipping line.
- The former fixing must be secure: if a former starts to become detached during dipping it can cause serious mechanical damage to the dipping line as well as the former.
- Formers will generally rotate at several stages in the overall dipping process. If any fixtures on the former mounting are screw fittings, ensure that the thread directions are such that the mounting will not tend to come unscrewed during these rotations. This may mean that double-sided plants may require a left-hand screw thread on one side of the line, and a right-hand thread on the other side.

It is not always possible, for various reasons, to dip with a full complement of formers on the line. If this situation arises on a chain plant, it is important that any gaps in the former complement are spread out evenly along the chain rather than concentrated in one place. When the formers enter the latex (or any other dipping tank on the line) the liquid level will rise owing to the displacement caused by the formers. If there is a long section of missing formers, this will mean that the displacement is less, hence the dip level will be lower. In an extreme situation this can lead to problems with, for example, beading. If the level control in the tank is controlled automatically, a drop in the liquid level will also trigger the addition of extra liquid, and the level may then be too high when a subsequent, fully loaded, section of the chain arrives. With a batch plant, where the formers are immersed to the correct depth either automatically or by eye, this problem will not arise.

11 Finishing operations for dipped latex products

11.1 Introduction

A freshly dipped rubber surface – especially a surface dipped from natural rubber (NR) or synthetic polyisoprene – is very tacky. The vast majority of dipped latex products dried straight from the dipping line cannot be handled easily. They will stick to themselves and any adjacent product. They will not be able to be picked up, tested, packed or undergo any other processing without further treatment to kill this surface tack. Such treatments are commonly called 'finishing', sometimes 'secondary manufacturing', and can take several forms. The most often used finishing treatments are the following:

- Halogenation (primarily chlorination).
- Powdering.
- Treatment with a silicone oil.
- Application of a polymer coating.

'Flocking', the application of short cotton (or other textile) fibres to the rubber surface before the latex has fully gelled to give essentially a textile, rather than a rubber, surface is described in Chapter 7.

11.2 Halogenation

All of the halogens (fluorine, chlorine, bromine and iodine) will react rapidly with the surface of NR or synthetic *cis*-1,4-polyisoprene, or any other elastomer which has residual olefinic unsaturation. Halogenation is primarily a surface effect, and will typically make the rubber surface harder and more slippery. Although the surface will feel smoother and 'silkier' (chlorination is often referred to as 'silking'), a microscopic examination will show that the surface has been slightly roughened. Of the halogens, chlorine is far and away the most widely used. A solution of bromine in water can also be used to remove surface tack, although there appears to be no advantage in using this over chlorine and a bromine treatment will discolour the surface of the rubber more than chlorine.

The reaction of chlorine with the rubber is usually thought of as a simple addition across residual olefinic double bonds, but in practice is much more complex [1] and can involve several different reactions, depending on variables such as pH, temperature, concentration, ionic strength and so on. Several active chemical species are generated when, for example, chlorine gas is dissolved in water, including molecular chlorine, hypochlorite ions, hypochlorous acid, chloride ions and hydrogen ions. If

https://doi.org/10.1515/9783110638097-011

the chlorinating solution is acidified, as is common practice, the number and type of active species in the solution is even more complex.

In a typical chlorination process, the dipped rubber article is exposed to a solution containing chlorine ions at ambient temperature. The chlorination can be carried out away from the dipping line (e.g., by tumbling in an industrial washing machine) or on the dipping line by passing the product, still on the former, through the chlorinating solution. Of course, chlorinating on-line will only treat one side of the product – usually the inside, as the product is often inverted on stripping. Chlorinating in a tumbler will treat both surfaces, although the greater exposure of the outer surface to the chlorinating liquor means that this surface usually becomes more heavily chlorinated than the inside. Products will sometimes be chlorinated both on-line and off-line, which allows different levels of chlorination to be applied on the inside and the outside. This can be of use when dipping gloves, to give a more slippery inside surface, which can assist donning the glove, whilst the more lightly chlorinated outer surface retains some measure of friction to help grip.

The chlorine can originate from several sources. For many years chlorination was carried out using a solution of chlorine gas in water. The chlorine would be stored in large drums and piped through automatic valves to blend the gas with water as the tumbler was filled. One of the problems with this – apart from the significant hazards in handling a highly poisonous gas – was that the concentration of chlorine in the finished solution was difficult to control with any degree of accuracy, causing variations in the level of surface modification. For this reason, most manufacturers have now switched to using aqueous solutions of readily available chlorine donors such as sodium hypochlorite (NaOCl) or sodium dichloroisocyanurate (Figure 11.1).

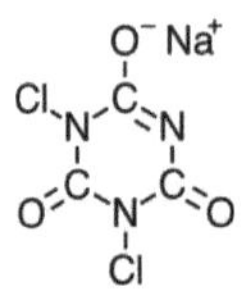

Figure 11.1: Sodium dichloroisocyanurate.

Solutions of both materials can also be used acidified, which generally gives a more rapid reaction. In off-line chlorination, the products are usually loaded into a tumbler, which is enclosed to prevent the escape of chlorine gas into the atmosphere. A water wash is often the first stage, which removes any residual materials from the dipping line, and then the chlorinating solution will be added. The chlorine concentration will depend upon several factors, including the material, the batch size and the type of tumbler, but for NR products will typically be in the range of 800 to 1,500 ppm (0.08 to 0.15%) of chlorine. The chlorination time will again vary, but is generally in the range of 3 to 10 min. Not surprisingly, the longer the time and the higher the chlorine concentration, the more intense the chlorination reaction will be. Chlorination of a NR surface can be fairly quick, but other polymers such as acrylonitrile-butadiene rubber or polychloroprene rubber can take longer. To end the chlorination, a neutralising solution (ammonia solution or a dilute solution of sodium thiosulfate) will be added, the tumbling continued for a few minutes, and then the solution discharged. The products are thoroughly washed with several changes of water, and then drained and dried by tumbling in a stream of hot air. The drying temperatures used on chlorinated articles are important. If too high a drying temperature is used, the product will discolour and give a yellow-brown tint, presumably caused by the loss of chlorine and the formation of conjugated diene species on the rubber surface. Generally, the temperature should be as low as practical, although many manufacturers strike a compromise between level of discolouration and speed and efficiency of drying. Chlorination with NaOCl is reported to be less efficient than with chlorine water and to cause more discolouration [1, 2].

On-line chlorination is carried out by including a chlorination dip tank into the manufacturing sequence. Owing to the need to control the chlorine fumes in the workspace, the concentration of chlorine in the solution will usually be lower and acidification may not be used. This will result in a lower level of chlorination and surface modification. Obviously, effective fume extraction over the chlorinating dip tank and associated rinse tanks will also be needed. Products chlorinated on-line will be well washed to stop the chlorination reactions and then progress through the subsequent manufacturing operations. As the chlorination reaction will have only affected one surface of the article, an additional finishing step will be required after stripping to kill the tack on the other surface.

A side effect of chlorination (which may or may not be beneficial, depending on the application and your viewpoint) is the destruction of other species that can be present on the surface of the article, such as residual accelerators and antioxidants. As rubber accelerators can be irritants, or promote allergic contact dermatitis (type IV hypersensitivity), halogenation can make dipped products such as gloves more dermatologically acceptable. On the other hand, rubber accelerators such as dithiocarbamates are powerful antidegradants, so reducing them may leave the rubber film more liable to oxidation. Halogenation has also been found to remove latex proteins from the glove surface, and will generally clean the surface of any residual powder from the dipping process.

Although chlorination has been claimed to increase the resistance of NR to synthetic detergent solutions [1], in general, it will reduce the physical properties of the rubber article, especially the ageing resistance. However, as mentioned above, chlorination is confined to the very surface, and any reduction of properties is usually negligible with thicker products, such as industrial gloves or catheters. With thin products such as condoms, chlorination will – unless very carefully controlled – reduce the physical properties to an unacceptable degree. For this reason, condoms are finished using one of the other methods.

Halogenation of dipped rubber articles can also be carried out with a solution of bromine in water. It is believed that the mechanism is similar to that of chlorine. The properties of a brominated surface are very similar to those of a chlorinated surface, although bromine is reported to stain the rubber more heavily than chlorine [1].

The effects of halogenation can be regarded as semi-permanent, and in many cases the reduction in tack will endure for the service life of the dipped latex article.

11.3 Powdering

Another common way of killing surface tack on dipped latex products is the application of powder. Unlike chlorination, powder is a temporary measure: as the powder is removed from the surface through wear, the surface tack can return. However, this is often not a problem with a single-use product, or if the service life of the product is expected to be relatively short.

The powder will usually be mixed with water to form a slurry, and the articles will be tumbled in this slurry for times ranging from a few minutes to several tens of minutes. The slurry will generally contain a low level of silicone oil (as an emulsion), which helps the powder adhere to the rubber surface. The amount of powder and silicone emulsion in the slurry can be critical as it controls the amount of powder actually picked up on the rubber surface. It might seem counterintuitive, but in the case of condoms, increasing the ratio of powder to silicone emulsion reduces the pick-up of powder on the condom; and conversely, reducing the proportion of powder results in more powder being picked up. This is attributed to the silicone oil providing the means whereby the powder can stick to the surface of the rubber. If there is a greater amount of silicone oil relative to powder, more powder is able to cling to the surface of the article.

Many different types of powders, or mixtures of powders, are used in the surface treatment of dipped latex articles. A list of the more commonly used powders is given below:

- Crosslinked maize starch (and other starch derivatives).
- Silica (silicone dioxide).
- Magnesium carbonate ($MgCO_3$).
- Calcium carbonate ($CaCO_3$).
- Talc.
- Lycopodium spores.
- Zinc stearate.
- Mica.

It should be noted that owing to current health and safety concerns, mica, talc and lycopodium spores cannot be used on medical articles such as surgical gloves, examination gloves and condoms. Any powder on such products will usually be found to be the more biologically acceptable $CaCO_3$, $MgCO_3$ or crosslinked maize starch.

Frequently, the slurry will be prepared as a concentrated dispersion and then diluted with water to give the concentration to be actually used.

After powdering, the products are drained and a centrifugation stage may be used to remove excess slurry residues, before they are dried by tumbling in a stream of hot air. It is common practice to reuse the slurry, but reuse to excess is not recommended. The water in the slurry will extract materials from the rubber that build up in the slurry, and subsequently contaminate successive batches of products. With NR products, any build-up of extractable latex proteins is an obvious concern. There is also the risk of bacterial growth in the slurry, especially if biodegradable materials such as starch powder are present or in the higher ambient temperatures found in tropical countries. Biocides can be added to inhibit microbial spoilage, although the level of any biocide on the finished product should be checked to ensure that there is no risk to the user, especially on those products where the user is subjected to long-term or repeated exposure. Many of the powders used in finishing slurries are alkaline, for example, $MgCO_3$ or crosslinked maize starch, which contains a small amount of magnesium oxide (MgO). Thus, the slurries themselves can be alkaline and many biocides are less effective at a higher pH. For these reasons, it is recommended that powder slurries are discarded and renewed at frequent intervals. There are obviously cost implications in discarding a part-used finishing slurry, but this can be mitigated by careful design of the process, e.g., optimising the concentration of ingredients and volume of slurry used. Utilising an already used slurry for a preliminary wash, then virgin slurry for a final wash is another way to give a clean final product whilst economising on slurry usage.

It is recommended that the slurry mixing and storage containers, as well as any pipework, are thoroughly cleaned at intervals, again to reduce the risk of microbial contamination; and the design of the slurry storage and feed should avoid 'dead' zones and difficult to clean areas where contamination may build up.

Apart from the effect on biocides, it has been found that the pH of the slurry can be important for the actual finishing process itself. The latices used for dipping are virtually always alkaline and the finished articles will frequently be pH sensitive. Exposing such products to an acid, or even neutral pH can change the surface characteristics, e.g., by changing any residual fatty acid soaps in the product to the free fatty acid. This can have the effect of increasing friction between the products whilst they are tumbling, despite the presence of silicone emulsion and powder, and such friction can create defects in the products. In one example, it was found that changing the pH of a starch-based condom slurry from around pH 10 to pH 7 caused a large increase in the incidence of holes under the bead of the condoms. When the pH was raised by the addition of a small amount of MgO, the holes disappeared.

A powder finish, however, cannot be used for certain products, such as gloves for use in the electronics industry, and it is preferred that surgical gloves are free from powder, because of the risk of powder remaining in the body after an intervention and causing postoperative complications such as granulomas. For these products, alternative finishing procedures must be used. For example, one such procedure has been patented [3] where the powder is dispersed in a polymeric binder, and the dispersion then used to coat the dipped article. The powder is thus fixed to the surface. The effectiveness of this type of finishing however, is not considered to be anything like as good as a standard powder finish or many of the polymer coatings.

11.4 Silicone oil

A rubber surface can also be rendered tack-free by applying a very thin coating of silicone oil. The oil is almost always used as a dilute emulsion, to avoid excessive amounts of silicone oil being applied. The products can be washed directly in a diluted silicone emulsion, then dried; or pieces of foam rubber or rags can be soaked with the emulsion and then tumbled with the products. An advantage of this latter method is that no subsequent drying is required, although the silicone will only coat the outside of the articles. If the inside has already been treated, by, for example, flocking of a glove or on-line chlorination, this can be an easy and cost-effective method to remove surface tack on the outside of the article. The effects of a silicone oil treatment are temporary, but as mentioned above, this is not a problem for single-use articles or those designed to have a limited service life. Products such as gloves or finger cots designed for use in the electronics industry cannot be finished with silicone oil, as contamination by silicone will give problems in today's highly sophisticated electronic industries.

11.5 Polymer coating

An alternative to chlorination to give a permanent modification of the rubber surface is to attach a polymer coating. In theory, at least this allows the surface characteristics to be tailored to suit the requirement. Such gloves have been around for well over 30 years now, following the launch of the Biogel® powder-free surgical glove in 1983. Powder is a complication on surgical gloves, as any powder entering an open wound during surgery can lead to postoperative complications, and separate washing treatments to clean these gloves are costly, time-consuming and not always very effective. The challenge with a surgical glove is not just to render the surface tack-free, but also to give an inner coating that is slippery enough to allow a close-fitting glove to be easily donned on a hand that has just been scrubbed, and can be dry, wet or any stage in between. The introduction of the Biogel® glove was the start of a wave of powder-free surgical gloves; some of them merely chlorinated, others carrying a polymer coating. However, attaching another polymer to the surface of a dipped latex article is not easy. The surface of the dipped article is flexible, elastic, and frequently of low modulus and hence easily distorted. In addition, the rubber surface can be contaminated with species such as sulfur, accelerator and antiozonant waxes, which may already have bloomed to the surface, and need to be removed before any polymer coating can be successfully applied. There is not space here to go into detail of the many different polymers which have been tried and are being used to coat latex dipped articles, and many of the formulations are patented and confidential. However, despite the technical challenges, many other effective coatings for gloves and other latex dipped articles have been developed since 1983.

11.6 Other approaches to finishing

Apart from the approaches described above, there are other ways that have been used from time to time to modify the surface of dipped latex products to meet various service conditions. These include applying a layer of wax to the surface. Antiozonant waxes work by blooming to the surface, thus providing a different surface layer. However, the blooming reaction is usually too slow to be of use in modifying the surface during the production process, but many waxes are available as, or can be made into, emulsions. Products can be tumbled in a wax emulsion in the same way that a silicone emulsion can be applied. Articles finished with a wax are similar in many ways to those finished with silicone, and like silicone, the surface modification is not permanent. Other materials that will bloom and remove surface tack when incorporated into the latex formulation are reported to be silicone oil emulsion or one of the polypropylene glycols [4]. It is claimed that levels of 1 to 2 phr can be incorporated into the formulation without adversely affecting the latex or the finished product [4].

Chemical modification of the rubber surface by means other than chlorination has been proposed, including epoxidation, plasma discharge, corona discharge, ultraviolet light and grafting of other polymeric entities, but at present none of these seems to have reached any commercial significance.

References

1. D.C. Blackley in *Polymer Latices: Science and Technology*, Volume 3, Chapman & Hall, London, UK, 1997.
2. *The Vanderbilt Latex Handbook*, Ed., R.F. Mausser, R.T. Vanderbilt Company Inc., Nowalk, CT, USA, 1987.
3. G.F. Stockum, inventor; Arbrook Inc., assignee; US4143109, 1979.
4. *Natural Rubber Science and Technology*, Ed., A.D. Roberts, Oxford University Press, Oxford, UK, 1988.

12 Troubleshooting in latex dipping

12.1 Introduction

All latex dipping processes will yield some scrap from time to time, even with a controlled and well-run production process. The key question here is 'what is done with the faulty product?' It is often said that one learns more from a failure than from a success. Failure can lead to investigation, and an investigation may uncover previously unknown details. Production scrap is no exception. An identification of the cause of the problem will lead to a solution being proposed. If carefully monitored when implemented it will soon be seen if the problem has been solved. If so, move on to the next problem and try and do the same for that: if not, examine the problem again, and see if another solution can be proposed and tried. This approach will be familiar to anyone who has been involved in 'firefighting', but is equally valid for continuous improvement programmes that should be part of the company's ongoing development programme. The key here is to set up a consistent and controlled method of defect analysis and troubleshooting, the subject of this chapter. In general, the most important (and frequently the most difficult) thing is to identify correctly the cause of the defect: once it has been identified, the action to be taken is usually obvious.

The pretty well universal response to a faulty dipped latex article – or anything else for that matter – is to look at it. This simple visual examination, if done correctly, can yield very useful data. The important words here are 'if done correctly'. It is very easy to get conflicting opinions – one person's bubble may be another person's foreign body inclusion. It is essential that those involved in the analysis of defects keep an open mind, and do not come to any conclusions in advance of the detailed examination. In order for a defect analysis programme to be of any use the analysis must be rigorous, controlled and informed.

A useful approach to process control and troubleshooting is given by Oakland [1], namely:

- No process without data collection.
- No data collection without analysis.
- No analysis without decision.
- No decision without action (which may include no action necessary).

A good knowledge of plant and process is important, as this will help both in the identification of possible causes of any defect and the most suitable corrective action to be applied. It is recommended that the dipping line operators are involved in any such defect analysis programme. Experienced operators are the ones who will know the plant best, and if properly trained will be able to pick up any signs that things are starting to go wrong. Mechanical dipping lines rarely fail catastrophically – there will

https://doi.org/10.1515/9783110638097-012

usually be some warning signs, and if recognised, potentially expensive production losses can be avoided.

12.2 Sampling

For successful troubleshooting, data will be needed. In addition, a good baseline of data is necessary, so that any deviations from normal can be recognised and quantified. For latex dipping processes it is recommended that product is sampled for in-process control checks as soon as possible, for example, immediately after the drying cycle, as close to the primary manufacture as possible. If anything is going wrong you need to know about it as soon as possible. Many latex dipping processes run at a high line speed, and it is important to recognise and fix problems quickly: a fault on a dipping line can produce scrap at a very impressive rate. So take samples out of the drier and check them. The number of samples will depend on the product, the batch size, the process and how much is known about the process. Sufficient samples should be taken so that there is a good chance of detecting any problems, but not so many that the testing becomes expensive and time consuming. The tests applied should be relevant to the product. For condoms and gloves, a simple water test will show if there are any holes, and a careful examination will show up any visual defects.

12.3 Examination of any defects

If any defect is found, it is recommended that the position is marked with an indelible pen – for example, some holes can be very small and difficult to find once the indication supplied by a water leak has been removed. The position should be recorded – preferably as accurately as possible, i.e., as '30 mm from the end', or '75 mm from bead' – rather than by a zone. This can show if there are any patterns in the position of any defects and help in subsequent analysis. It is also helpful to know if the surface being examined is the inside (former contacting) or outside surface – an examination of the bead (if present) or any patterning can help here. With a condom or glove it will often be possible to determine which is the inside or outside by looking at the bead. If one is looking at the outside (as dipped) surface, the film close to the bead will appear smooth and even: if the product has been inverted so that the inside or former contacting surface is being inspected, the film next to the bead will appear to be puckered (Figures 12.1 and 12.2).

Similarly, it is helpful to know the orientation of the product, especially if only a portion of the article is being examined.

Visual examination is best carried out using a stereo microscope. Hand-held or stand-mounted magnifying glasses are useful, but a stereo microscope with an

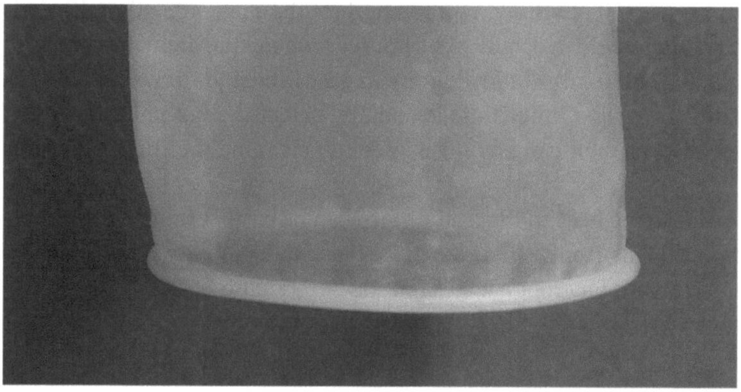

Figure 12.1: In this photograph, the exterior of the condom is the outside as dipped. Note that the film beside the bead is undistorted.

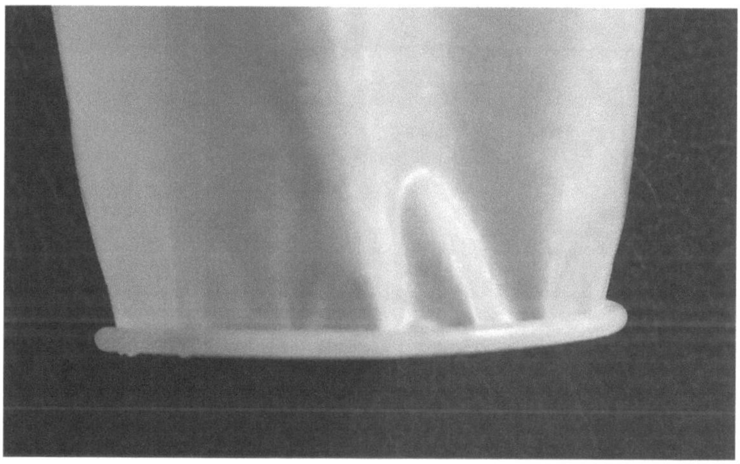

Figure 12.2: In this photograph, the condom has been inverted, so that the exterior of the condom is the inside (former contacting) surface as dipped. Note that the film beside the bead is puckered.

adjustable and directional light source will reveal so much more detail, and can be coupled with a camera to record the images. High magnification is not necessary – in fact, using a high magnification can frequently be more of a hindrance as the depth of field becomes very small and the very limited area that will be visible means that it is easy to lose context. Also of importance is the working distance – the distance between the objective lens and the sample. If this is too small it will be very difficult to access the sample, for example, to mark points of interest or clean the surface. There will also be the risk of the housing of the objective lens masking the illumination. Magnifications of approximately 15 to 50X are generally the most useful. The light source is, if anything, more important than magnification. It needs to be adequately bright, and a 'cold' light source, where the light is conveyed through a flexible fibre-optic tube, allowing the angle of the light to be varied so as to highlight the details of the rubber surface and any defects. A soft paintbrush and a solvent such as ethyl alcohol can be of use in gently cleaning the surface to show more detail.

The sample needs to be held so that the surface can be easily examined. A simple sample holder can be made from a short length of plastic tube, for example, some 30 to 40 mm in diameter, with any rough edges rounded off (Figure 12.3).

Figure 12.3: A simple sample holder can be made from a plastic tube. In this example, a cap to secure the sample can also be seen.

The sample can be placed over it and secured with an elastic band or a specially made cap. In some cases, for example, where the point of interest is located next to the bead, a tube that can be held horizontally will be more convenient. One thing to remember is that the sample must not be stretched when held for examination, as this will distort and may damage any defect, and lead to an incorrect diagnosis. It is very important that the examination is carried out in a controlled and rigorous fashion.

It is essential for the successful and meaningful identification of a problem that a consistent approach is always followed. Develop a standard method for the examination, validate it and train reliable staff to carry it out. It will help if photographs of various types of defect are available to act as controls, and it is recommended that staff are challenged and retrained at regular intervals. Defect analysis is a powerful tool for improvement, but only if carried out properly. A wrong diagnosis will lead to wasted effort, and perhaps more importantly, may lead to the process being abandoned. Above all, if the cause of the defect is not known, that must be recorded. A guess will usually not help, and can be counterproductive. There will be many occasions when the cause of the defect cannot be pinpointed, but as the operators become more experienced in defect analysis the proportion of correct diagnoses will increase and the whole troubleshooting process will become more effective. More information can be found in the article by Hill [2].

12.4 Collection of data

Accurate data are essential, not only for problem solving, but also for the general maintenance of a good level of quality. However, only gather the data that will be used. The author has seen too many cases where a property is measured because it can be measured, rather than because the results will be useful. These results will then be stored away in a filing cabinet or on a computer hard disc, never to be consulted again. Remember that measurements and data collection cost time and money. Unnecessary information can obscure the real picture and can lead to data overload. For example, some (not all) of the parameters available for measurement on a dipping line are shown in the list below. Measuring all of these as a routine will rapidly swamp the system with data, many of them irrelevant and duplicated.

Some of the parameters that can be measured on a latex dipping line:

- Beading brush speed
- Coagulant strength
- Coagulant temperature
- Dipping line speed
- Drying oven temperatures
- Film temperature
- Latex alkalinity
- Latex degree of precure
- Latex mechanical stability time
- Latex pH
- Latex temperature
- Latex total solids content (TSC)
- Latex viscosity
- Latex agitation speed
- Leach composition
- Leach temperature
- Preheat bath temperature

- Stripping set-up
- Stripping solution composition
- Vulcanising oven temperatures and many others...

So it is very important to select the correct data, both from the dipping line and from the actual product to use in quality monitoring.

The way in which the data are gathered and recorded is important. Try to record the data in a way that preserves the nature of the data. For example, a tick to show that a parameter is within specification and a cross to show that it is out of specification means that any subsequent analysis will be limited. On the other hand, numerical values can be graphed, can show trends and will indicate whether the result is close to the limit, or well within the specification range.

12.5 Corrective action

Once a problem has been noticed, examined and identified, corrective and preventive actions will need to be applied. Very often these will be obvious. If a former is misaligned and hitting part of the dipping line, remove or realign the former. If the beading is ragged or serrated, go and check out the beading set-up and film temperatures at beading. However, there will be times when further experimentation will be needed to identify, correct and thus prevent a defect. In these cases, the following hierarchy of actions is proposed.

First consider making a 'physical change'. Physical changes include things such as:
- Increasing or decreasing the line speed.
- Increasing or decreasing oven temperatures.
- Using different combinations of heating.
- Increasing or decreasing beading brush speeds or brush alignment.

Physical changes are usually quick and easy to do, with minimal extra cost and the effects are generally seen quickly. Importantly, they can be easily reversed if the change does not solve the problem.

Should a physical change not work, then consider making a 'mechanical change'. Mechanical changes can include:
- Adding extra heating or cooling stages.
- Changing exit or entrance speeds or cams.
- Adding extra beading stages.
- Installing extra equipment.

Mechanical changes are relatively simple to carry out and usually at moderate cost. They can easily be reversed if necessary, and the effects will be seen quickly.

Only if physical or mechanical changes do not work should 'chemical changes' be considered. These can include:
- Changing the latex formulation.
- Adding extra chemicals to the various dipping tanks on the production line.
- Changing the materials in the finishing processes.

Chemical changes cannot be reversed, you cannot take out a chemical you have added to the latex. The full effect of chemical changes may not be seen for some time, and there may be unforeseen effects. For example, adding an extra material may solve the problem as seen at production, but further down the line there may be problems with ageing or allergy. In addition, chemical changes on a medical device may need additional and expensive studies such as clinical trials and toxicity testing. For these reasons, it is recommended that chemical changes are only applied as a last resort.

Note that change control is mandatory in quality management systems [International Organization for Standardization (ISO) 9001:2015, clause 8.5.6; and ISO 13485:2016, clause 7.3.9], and in these cases there must be a recognised and documented system to carry out, verify, validate and record any changes.

12.6 Some types of defect and suggested corrective and preventive actions

12.6.1 Mechanical damage

Mechanical damage can be divided into damage that occurs on the dipping plant and that which is caused after dipping. In general, it is an easy form of damage to identify. On the plant, the main problems will be knocks and rubs, and sometimes damage caused at stripping. Formers are generally mounted as close together as possible, to maximise the output from the dipping plant. This means that if there is any misalignment, or precession when rotating, the formers can touch or 'knock', damaging the latex film on the former. This type of damage is generally seen towards the end of the article, for example, around the shoulder of a condom or fingers of a glove. Knocks and rubs are usually large. The solution is, unsurprisingly, to ensure that the spinner shaft is straight, the bearings are in good condition and that the former mounting is properly aligned. A similar defect on chain plants is a 'rub', where the former contacts part of the dipping machine as is progresses through the cycle. A rub can often be created if there is any dripping of latex from the former at, for example, the entrance to a drying oven, or any other station close to the exit from the dipper where the former passes. This latex may build up sufficiently to contact the latex film on the former as it passes.

A less obvious type of mechanical damage is a scratch, which is caused by something hard and/or sharp scratching over the rubber film, and identifying the cause can be difficult. The orientation of the scratch can help in this area. Often the scratches look like a series of 'v's, and where this is seen the point of the 'v' indicates the direction of initiation of the scratch. For example, scratches consistently aligned along the axis of a condom, in the direction bead to teat, point to damage that may be occurring at stripping. Scratches which appear similar, but are aligned randomly relative to the product axis, can be caused by the washing machine or tumble drier, used in the slurry finishing. Small (sometimes very small) protrusions such as exposed welded seams or the ends of wire mesh can damage the articles as they tumble around. These protrusions can be very difficult to find, but changing the washing machine or tumble drier to another will confirm whether or not it is to blame.

Some examples of scratches are shown in Figure 12.4.

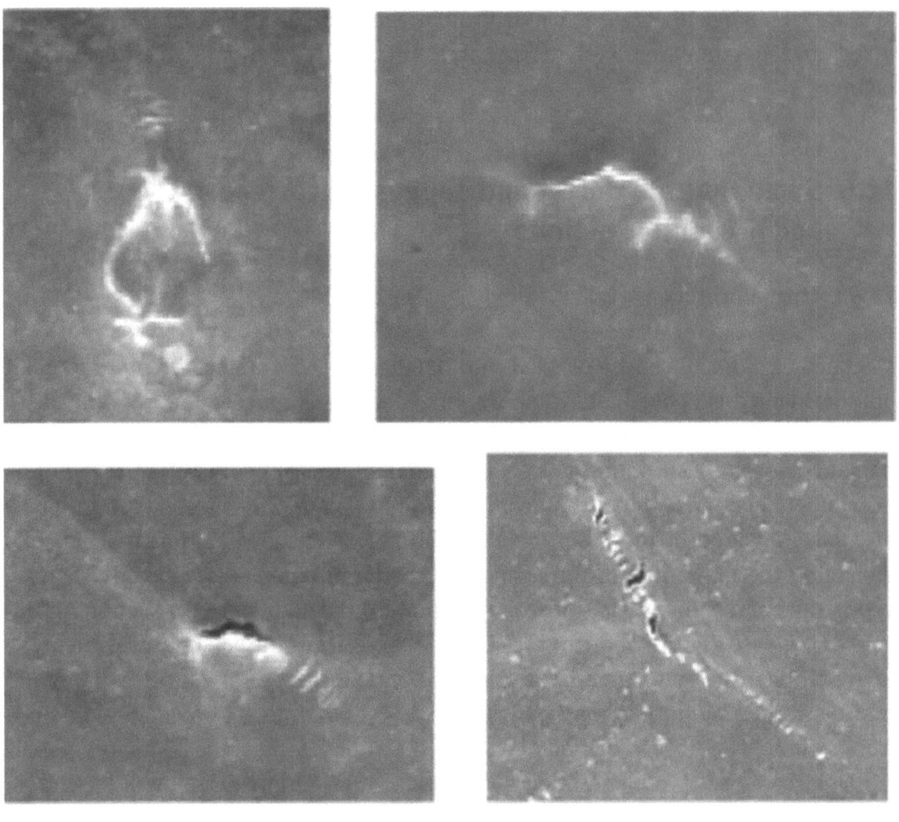

Figure 12.4: Examples of holes caused by scratches.

12.6.2 Other causes of holes

As well as holes caused by mechanical damage to the film, other causes can also result in holes. On gloves, webbing (either of the coagulant or latex) can give rise to holes in one or more of the different layers. These holes are always found in the crotch between the thumb and finger, or in the crotches between the fingers. They are caused by a film of coagulant or latex which spans the gap between the fingers, or fingers and thumb, as the former is withdrawn. This film then breaks and leaves an uneven film of latex or coagulant, which can then cause a hole or defect. There are effective antiwebbing liquids, such as medium chain alcohols or silicone emulsions, and also various proprietary antiwebbing agents of undisclosed formulation. An example of a crotch defect – in this case a 'half-hole', that is a hole in one of the two layers – can be seen in Figure 12.5. The glove has been inflated to show the defect more clearly.

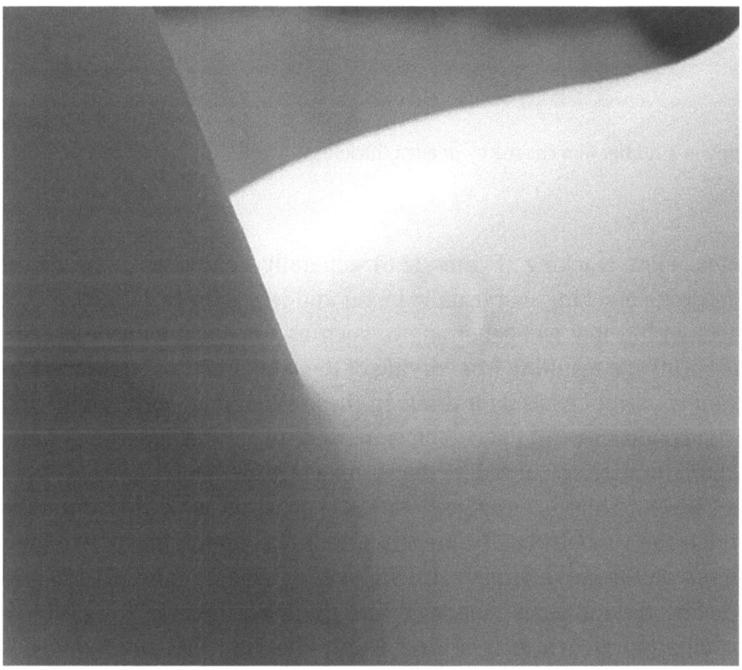

Figure 12.5: An example of a defect in the crotch of a houseware glove.

Where there are thick areas of latex adjacent to thin areas, often seen on the fingertips of a glove or around the teat end of a condom, the differential rate of drying and associated shrinkage can also lead to cracking, which in some instances can

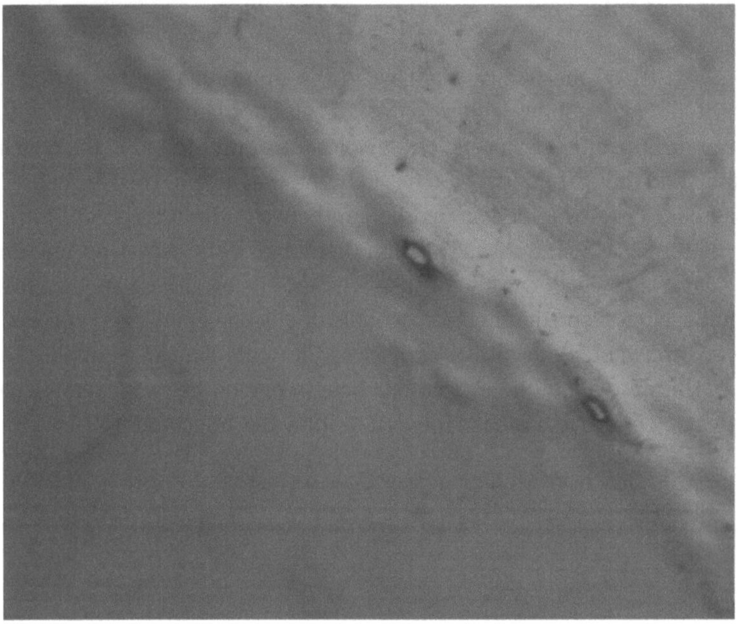

Figure 12.6: Cracking in a rubber film caused by uneven thickness.

give rise to holes. Film cracking (Figure 12.6) generally comprises a swarm of cracks, similar to ozone cracking, but usually larger and less sharply defined.

On elastomers with a low wet-gel strength, such as some acrylonitrile-butadiene rubbers, the shrinkage resulting from drying of the latex film can stress the material to the extent where it cracks as it dries. In these situations, very careful and progressive drying conditions will have to be employed, to get the material to build strength faster than the stresses caused by the shrinkage grow.

A particular cause of holes in supported gloves is penetration of fibres from the fabric liner through the rubber film. To prevent this it is common practice to pass the liners, mounted on the glove formers, briefly over a flame, to burn off any protruding fibres, before the former is immersed into the latex. If there is a problem with holes on supported gloves, it is always worth checking that the flaming is being carried out correctly.

12.6.3 Coagulum and cream

Coagulum can be very difficult to see. It is effectively looking at a piece of rubber in a rubber matrix, and unless the particle size of the coagulum is very large it will be very difficult to detect, although it may be more visible on a partially dry film where it can show up as white specks as the film becomes transparent. It is better to try and confirm any suspicions about coagulum by filtering a sample of the latex through a fine mesh before drawing any conclusions. When the mesh is rinsed with ammoniated water any coagulum retained on the mesh will become apparent.

Cream, or areas of thick latex on the former, on the other hand are easy to identify and resolve. On a chain plant, the latex dippers can be difficult to agitate effectively, as the shear-sensitive nature of the latex effectively precludes any rapid agitation. This will be less of a problem on a batch plant. As a result of inadequate agitation, the rubber particles may rise to the surface and give a thicker surface layer, or 'cream'. In addition, lack of agitation may allow a surface skin to form, which can be picked up onto the former. Good dipper design and a programme of regular dipper maintenance will ensure that these problems are resolved.

12.6.4 Bubbles

Bubbles can be very easy to identify, they look like bubbles. The cause may be easy to find: small bubbles randomly distributed, shortly after an intervention such as filling the dip tank, and which diminish with time, point to air having been introduced during the intervention. Alternatively, bubbles that do not diminish, or even increase with time can be the result of a leaky seal on a circulation pump drawing air into the latex or excessive agitation. A cluster of bubbles at the teat of a condom, finger tip of a glove, or any other thicker area of latex, is usually caused by overdrying – the heat has been turned up to dry the thicker film of latex, which has caused it to skin over, trapping wet latex underneath. If the temperature rises too high, volatile materials, such as water, can form bubbles in this wet area sandwiched between the dried rubber surface and the glass.

Examples of these overdrying bubbles can be seen Figure 12.7.

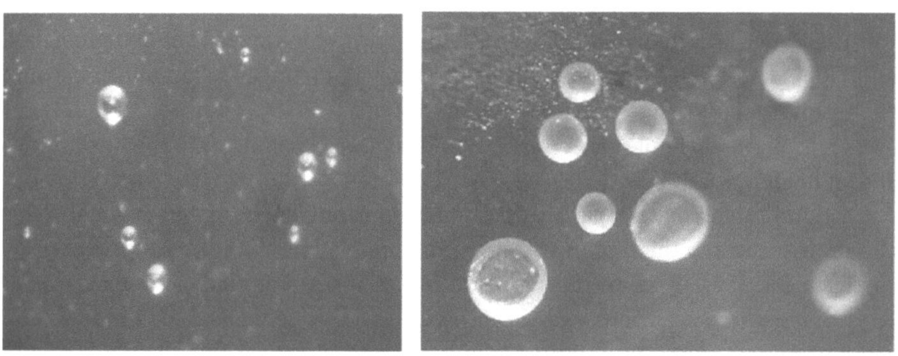

Figure 12.7: Examples of overdrying bubbles within the film.

As mentioned in Chapter 7, bubbles can at times be formed on the very end of the former as it exits the latex. These will frequently burst immediately on heating, but if not this problem can often be resolved by changing the rheology of the system slightly – a marginally faster exit speed coupled with a slightly lower latex viscosity; or a marginally slower exit speed/higher latex viscosity should be tried. Alternatively, a gentle jet of air blowing over the end of the former as it leaves the latex surface can help. Examples of bubbles in a latex film can be seen in Figure 12.8.

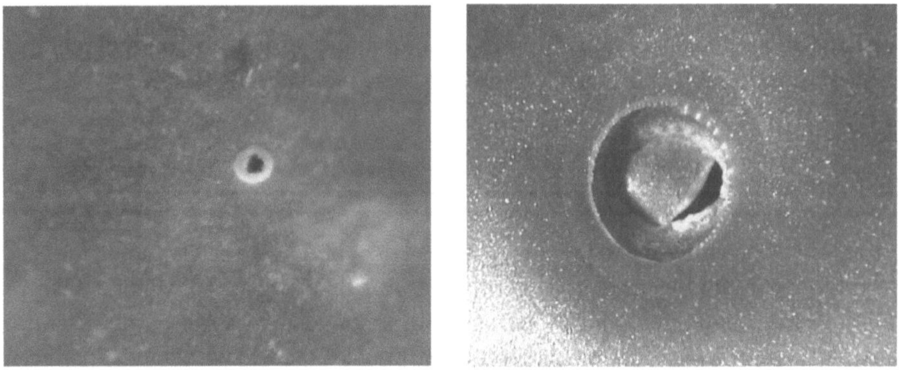

Figure 12.8: Holes resulting from bubbles in the rubber film.

12.6.5 Foreign bodies

Foreign bodies cover both ends of the identification spectrum. Sometimes they can be very easy to identify, as the rubber film retains the shape of the solid particle that was (and often still is) in the film. Other foreign bodies can be small, hard to see, and at times are difficult to distinguish from bubbles. An example of a foreign body can be seen in Figure 12.9.

The cause is often inadequate milling and/or filtering of solid ingredients, although some foreign bodies, such as fibres or filaments of rolled-up rubber film can result from cleaning the dippers and surrounds.

An example of a hole caused by a fibre being incorporated into the latex film is shown in Figure 12.10.

Natural rubber (NR) in particular can have a strong odour of ammonia, and this can draw insects to the dipping tanks – ammonia is one of the gases produced by rotting meat. It is therefore not unusual to see insects around, and even in, NR latex. Adequate precautions against insects, including screens at all windows and other openings, screens around the dip tanks and electrical fly killers are recommended, especially in tropical countries. As with all of these types of defects, once

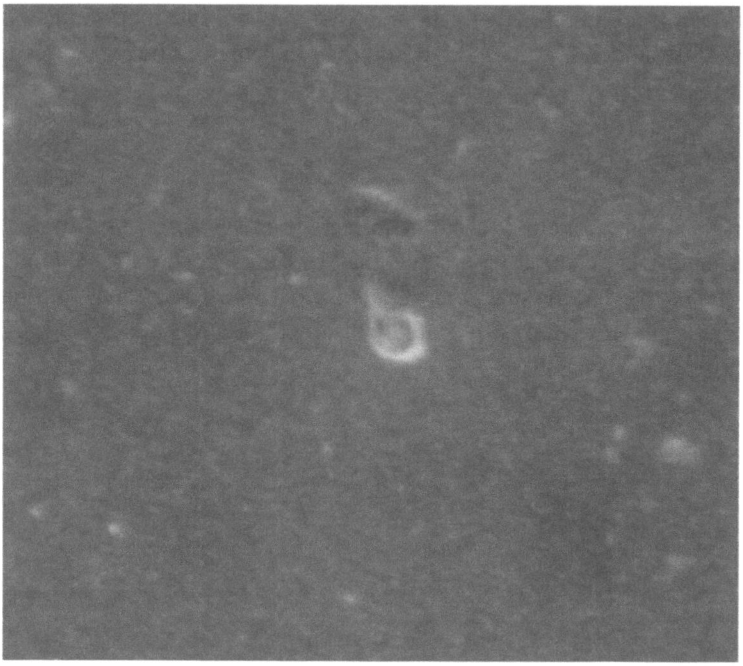

Figure 12.9: Particulate foreign body in the rubber film.

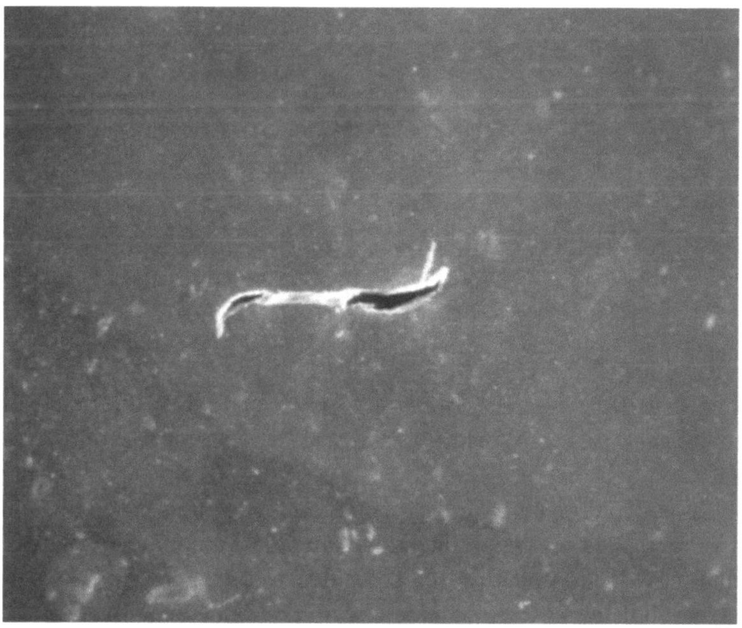

Figure 12.10: Hole caused by a cotton fibre trapped in the rubberfilm.

a cause has been proposed, corrective action can be applied and the production monitored to see if the action is successful. If not, the problem must be reexamined.

12.6.6 Damaged formers

Even though the materials generally used for formers (borosilicate glass or porcelain) are tough materials, a former can sometimes be damaged by, for example, contact with parts of the dipping machine frame or other formers. Damage on the end of a condom former, caused by it repeatedly hitting the frame on a chain plant is shown in Figure 12.11.

Figure 12.11: Damage at the end of a condom former.

It is also possible that defects on the former can be created when the former is made, e.g., bubbles may be created close to the surface when a glass former is shaped (Figure 12.12).

These bubbles may subsequently break open, and any scratches or chips on the former will be replicated on the product, and can create a point of weakness. Routine examination will soon pick up the coincidence of position and appearance of the damage, but the next problem is to find the defective former or formers! A modern dipping plant can have several thousand formers mounted on it, and finding one or two damaged formers can be a daunting task. For this reason

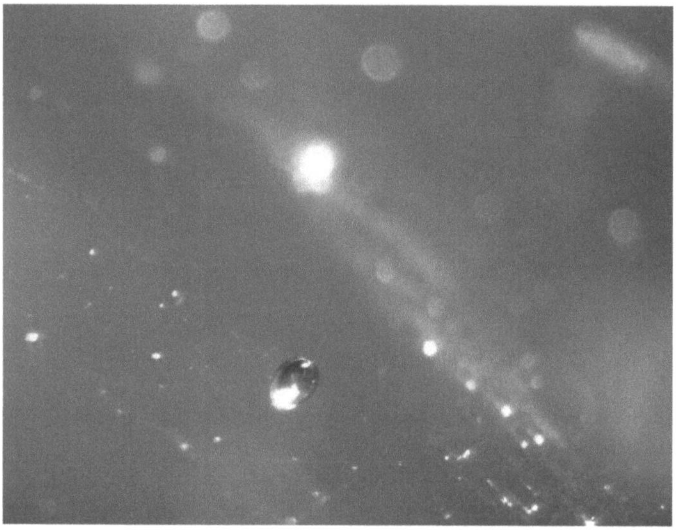

Figure 12.12: Bubble just below the surface of a glass condom former.

many companies will have numbers engraved on the formers that allow any defective ones to be rapidly identified. However, it is important that any engraving or other means of numbering the formers does not, in itself, create the potential for flaws.

12.6.7 Stuck work (also known as 'Flats' or 'Stickies')

Occasionally, the finishing process may not be 100% effective at removing the surface tack, and some condoms or gloves may have areas where the film is stuck to itself. These areas can range from very large to very small, and the stuck surfaces can be on the inside or outside of the product – or both. If the surfaces are separated after sticking, a hole or tear may be caused. Such holes can be difficult to assign at times, although if the hole is adjacent to a piece of rubber stuck to the surface, of the same shape and size, the task is made easier. Figure 12.13 shows a hole caused when a condom stuck to itself was pulled apart. The piece pulled out of the hole was located near to the hole, still stuck to the other part of the condom.

The cause will usually be an incorrect finishing process, and with a powder finish, for example, attention should be paid to the age and composition of the finishing slurry – the concentration and ratio of powder, silicone emulsion and surfactant. For chlorinated products, the problem will often be an incorrect pH or low concentration of the chlorinating agent.

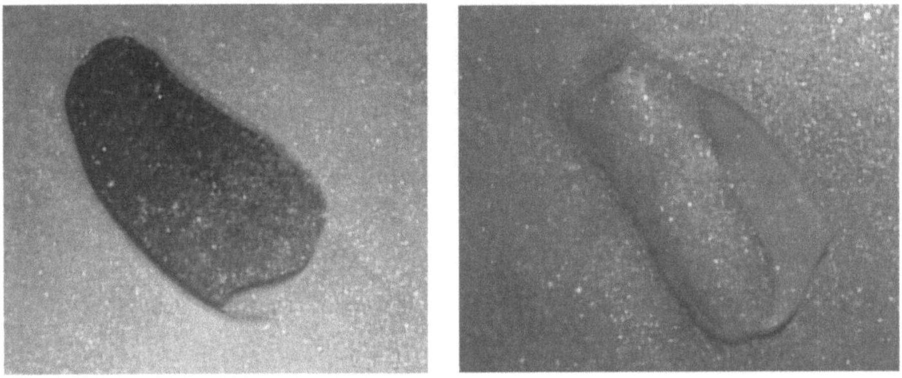

Figure 12.13: Hole caused by internal stuck. The right-hand photograph shows the piece of rubber pulled out of the hole, still stuck to the rubber film.

12.6.8 Unwetted marks (also known as 'Fish Eyes')

Unwetted marks are always caused by contamination, usually of silicone oil or lubricant. They manifest themselves as circular areas, which can be quite large, in the latex film, seen most clearly as the film starts to dry. The contaminant has prevented the latex wetting the substrate properly, and as the latex starts to dry, it will pull away from the contaminated area.

There can be several causes of the contamination. Oil from moving parts of the dipping line may have found its way into the latex or onto the formers. Silicone emulsion from a powder-finishing area may be the culprit, and even contamination of storage tanks or delivery tankers has been known to be responsible. If the cause of contamination has been identified and removed it is sometimes possible to remove the contamination by continuing to dip (and scrap the bad product!) until the contaminant has been cleared. Another action, which sometimes works, is to turn off the agitation, let the contaminant come to the surface of the latex (the usual culprit, oil, is lighter than latex) and cover the surface of the latex with absorbent paper to remove the material on the surface. This can be repeated several times if necessary. However, in many cases it will be necessary to scrap the latex, clean the dip tank (and any other affected equipment) and start again.

12.6.9 Bead defects

There are many different types of bead defects, some serious, some less so. These include 'serrated' beads, where the surface of the material on the bead is rough, which is usually caused by the film being too hot or too dry as the bead is formed.

'Knotted' beads, where there are one or more pronounced thicker lumps on the bead. Typical causes of knotted beads are overdrying, so that the film is difficult to release from the former. A worn beading brush may also give rise to knotted beads.

'Figure-8 beads', where the relaxed bead has assumed the figure-of-eight shape. This is caused by the beading brushes or rollers not being correctly adjusted, so that strain is built into the bead as it is formed, as described in Chapter 7. When removed from the support of the former, the strain relaxes, pulling the bead into the characteristic shape.

'Blown' or 'blistered' beads are the result of the temperature after beading being raised too high too quickly, whilst there is still moisture present in the bead. If this water boils, it can cause the bead to bubble and blister.

Holes under the bead can have several causes, usually connected with incorrect drying. They can often be found together with knotted beads. Another cause can be inadequate tack of the film when the bead is formed – again usually the result of overdrying – leading to the bead partially rolling back, leaving a weak area immediately under the bead.

Figures 12.14 and 12.15 show examples of holes under the bead.

Figure 12.14: Example of holes under the bead of a NR condom.

Note that on Figure 12.15 the bead has rolled back a little, leaving a thin area immediately under the bead. It is worth noting that a low pH (say 7–8) in the washing slurry at finishing, which tends to increase the friction of NR products, has also been found to generate holes under the bead. If holes under the bead seem to be appearing after the finishing process, it might be worth checking the pH of the

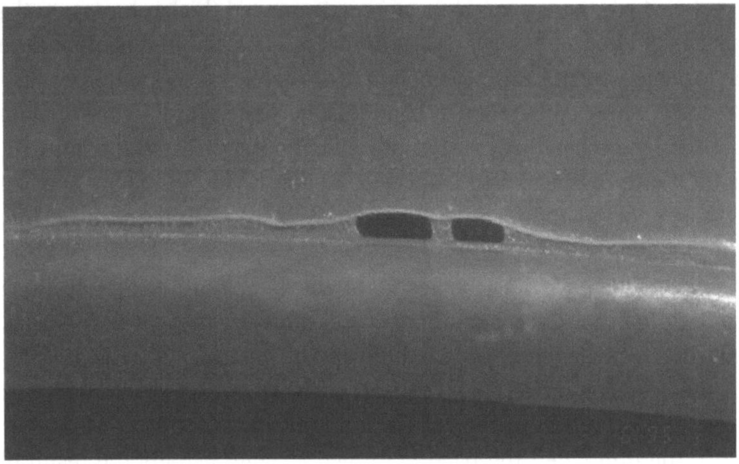

Figure 12.15: Example of holes under the bead of a NR condom.

finishing slurry, and possibly increasing it by the addition of, say, a little magnesium carbonate.

As can be seen, many of the bead problems are related to drying of the latex film, and that should normally be the first parameter to be checked if there are significant numbers of problems with the beads of a beaded product.

12.6.10 Discolouration

If chlorinated rubber products, especially NR articles, are heated for too long, or at too high a temperature, they will discolour as the chlorinated surface of the rubber begins to degrade. This can occur quite quickly at temperatures over 70 °C, and will be very noticeable with pale-coloured articles. The only remedy is to use the minimum level of chlorination possible and the lowest drying temperatures commensurate with production constraints. This will not eliminate the problem, but can reduce it to acceptable levels.

Some pigments can also change colour at high drying temperatures, and such temperatures can also promote colour bleeding of certain pigments.

12.6.11 Double imaging

Typically, latex dipping will take place with the latex at 40 to 50% TSC. In other words, at least half, and often more, of the mass of latex picked up is water, and will be lost as it dries. This leads to shrinkage, which will increase the risk of cracking. The shrinkage tries to occur in three dimensions: the film will become thinner, and it will try to contract around and along the former. If there is insufficient friction between the film and the former surface, such as may occur if there is a film of water between the rubber and the former, or if there is an excess of the powder added as a parting aid, the rubber film can slide down the former. This will not only affect the length of the finished article, but if the former carries a pattern, as is often seen with household or industrial gloves, then double or multiple imaging of the pattern may result. This can be a difficult problem to overcome, although the following actions may help:

- Make the drying stages as gentle as possible (within production constraints).
- Use as low a leach water temperature as possible, or even better, leach off-line.

12.7 Conclusion

This chapter has briefly covered some of the possible causes of damage that may be encountered during the production of dipped latex articles. There will be others, many of which will be specific to particular production machinery and processes. Nevertheless, the principles described have wide application. A good knowledge of the production process – being familiar with what is happening to the product at every stage during production – is essential for the successful application of this technique, and for this reason well trained and experienced senior operators or supervisors are the obvious choice to implement it. Good record keeping is also strongly recommended, as a good database will greatly aid in both the identification and correction of any damage. Note that it is possible that the overall level of holes or other damage detected may not change from day-to-day, but if the cause of the damage changes, then that could be an early sign of greater problems on the way. Having a tool to recognise such a change will help to maintain quality. No matter how high the quality of the product, there will always be room for improvement, and a well-planned and well-implemented scheme for the analysis and categorisation of damage can only lead to better quality, higher yields and lower costs.

References

1. J.S. Oakland in *Statistical Process Control*, 6th Edition, Routledge, Abingdon, Oxford, UK, 2008.
2. D.M. Hill in *Proceedings of the Latex and Synthetic Polymer Dispersions Conference*, Madrid, Spain, 2008, Paper No.15.

13 Quality control in latex dipping

13.1 Introduction

Many products dipped from latices are medical devices, and as such, are regulated. It is a fact of modern life that regulatory requirements tend to increase, not decrease over time. As a consequence, the costs of implementing these requirements similarly increase. Add in the fact that raw material costs, in general, increase and environmental constraints similarly tend to increase, manufacturers dipping such products find themselves faced with cost increases coming at them from several directions. How can manufacturers cope with these pressures?

An obvious answer is 'improve quality'. An equally obvious response is 'how?' – this is where quality control techniques can come in. An effective quality control system, applied throughout the whole manufacturing process, from design to dispatch, will improve manufacturing, improve quality and reduce waste. It will help to clarify the real, underlying cause of any failures and prevent the design and manufacture of defective products. The basic quality control concepts span all manufacturing industries. However, it is recommended that the quality targets remain realistic. Perfection will almost certainly be expensive and may very well not be needed. Specifications and tolerance limits should be appropriate to the function of the product. It would not be sensible to apply the same criterion for hole levels to toy balloons as for condoms or surgical gloves.

Nowadays there are many quality control techniques, of varying degrees of sophistication, although this chapter will only look at those that can be applied easily and effectively in a latex dipping environment. This chapter does not aim to be a comprehensive treatment of quality control methods, but hopefully will serve to introduce the reader to the concepts and prompt further study where appropriate.

Quality control methods will, invariably, rely upon the collection and interpretation of data, and how these data relate to the required values. For this purpose, it is much easier to use graphical methods, as they can provide an immediate visual indication, not only of the current values but also of any trends. Graphical methods can include:
- Histograms
- Scatter diagrams
- Run charts
- Mean and range
- Attributes charts
- Pareto diagrams
- Moving average charts
- Cumulative sum (cusum) charts

https://doi.org/10.1515/9783110638097-013

Modern techniques of industrial quality control can be traced back to Walter Shewhart and the work he did at Bell Telephone Laboratories, Inc., in the 1920s. For this reason, many of the graphical techniques referred to above are known as 'Shewhart charts'. However, before we look at quality control techniques it will be useful if we define exactly what is meant by 'quality'. Several definitions could be applied, but in the field of latex dipping, the following could be appropriate:

- Fitness for purpose
- Conforming to the agreed/required specification
- Satisfying customer needs

In the past, it was not unusual for a manufacturer to produce the dipped latex articles, then apply an inspection routine to determine which of the products met the required 'quality' criteria. Those that passed were forwarded to be sold, those that failed would be either scrapped or returned for rework; but all of this costs money. It is far better to incorporate checks that help to get the product right at the point of manufacture, and this is where quality management systems, often going under the heading of statistical process control (SPC), can help.

An SPC system will require the application of statistics (as implied by the name) although there are effective procedures that can be handled with the minimum of mathematical knowledge.

Inherent in these control procedures is the fact that the operators need to know what the product specifications are, how to check them and what to do if the product does not comply. There are many sources of information on SPC and quality control available, but one that is recommended is the comprehensive and readable book '*Statistical Process Control*' by Oakland [1]. These techniques are also covered by a range of International Organization for Standardization (ISO) standards [2–7].

13.2 Analysis of data

In any manufacturing process there will be variation. The aim is to keep these variations as small as possible so that, to all appearances, the products are all identical. Furthermore, the properties of the products – at least the critical performance properties – should be maintained well inside the required specification limits.

SPC uses statistical tools to monitor the outputs from the production process to detect significant variations from the specification before any non-conforming products are produced, and hence eliminate or minimise waste.

Production will be subject to two types of variation, known as 'common' or 'non-assignable' causes, and 'special' or 'assignable' causes. Common causes are random or chance variations in the process, and can result from a variety of influences outside of the manufacturer's control. These could include changes in season, changes in atmospheric pressure, the inbuilt precision within the manufacturing or test equipment and so on. When only common causes are present in a manufacturing process, the process can be considered to be under control.

The opposite of common causes, the special, or assignable causes, are those that can arise from errors in the process. These can be such things as incorrect machine set-up, machine wear and tear, inadequate operator training, incorrect raw materials and so on. The aim of SPC is to identify those variations that are normal to the process, that is, those that result from random variations outside of the manufacturer's control; and those arising from assignable causes which once identified, can be addressed and corrected. If carried out correctly, monitoring the outputs from the production process will detect significant variations before any non-conforming products are made. This can be carried out through the following three phases:

1. Understanding the process and specification limits.
2. Identifying the special causes of variation and eliminating them.
3. Monitoring the process through the use of suitable control charts so as to detect any changes and initiate corrective action.

13.3 Data considerations – variables data or attributes data?

Test data from latex manufacturing can take two forms: attributes data and variables data. It is important to be able to distinguish between the two, as the manner in which the data are treated differs. Variables data are, as the name suggests, variable. They can take any value on a continuous scale, and fine levels of intermediate values. Examples of variables data are tensile strength, length, weight, concentration, pH and so on. Attributes data, by comparison, are discrete and can only take certain defined values. Examples are defects and the number of products passing or failing the inspection criteria. One cannot have 1.34 holes in a condom or glove, or 3.47 products failing to meet the test requirements.

It should also be noted that many of the mathematical treatments of the variables data in SPC assume that the results are normally distributed. This is very often not the case with real data from dipped latex products. For example, burst data on condoms are frequently negatively skewed [that is, the distribution has a longer 'tail' on the left-hand side (lower number)] e.g., Figure 13.1.

The opposite is a positively skewed distribution, where the tail of the distribution extends to the right-hand side of the graph.

In some cases this deviation from normality is small, and can conveniently be ignored, but there are other situations where lack of normality can be important. Fortunately, there is a theory called the 'central limit theorem', which states that, even if the individual results are not normally distributed, the distribution of the mean of the results will tend to normality. The greater the number of replicates

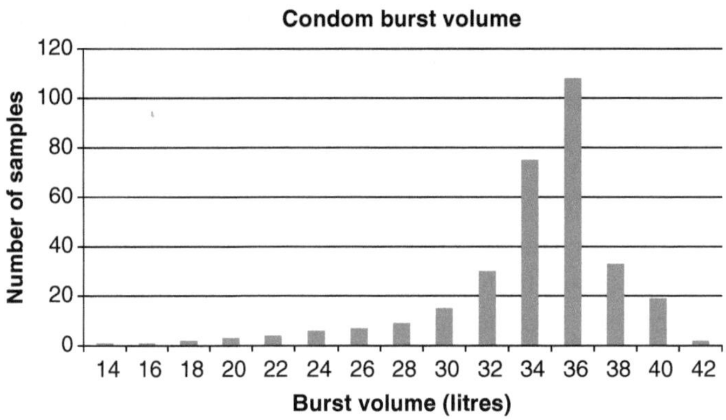

Figure 13.1: Example of a negatively skewed distribution.

used to calculate the mean, the closer the distribution will be to a normal distribution. Even with a small number of replicates, such as four, the distribution of the mean will be close to normality. For this reason, many SPC processes and charts will work with the mean of a small number of test samples.

13.4 Control charts

Control charts are an excellent way of monitoring the quality of production, and can show not only the current and previous situations, but, importantly, any trends. Several different types of control charts can be used, and interested readers are directed to the many different texts, for example, the book by Oakland [1] for more information. However, as noted above, test data can be gathered as variables data or attributes data, and these require different treatments.

13.4.1 Control charts for variables

Simplest of all is the 'run chart', where the test results are plotted on a chart against the date, time or sample number. Run charts give an immediate picture of the process, and it can easily be seen if the results are trending up, trending down, fluctuating or steady. The specification limits can be marked on the chart so results approaching the limit can readily be seen. A run chart gives an immediate picture but more information can be gained from more sophisticated charts, such as the mean and range of the samples given in the charts. Here, the mean of the samples is plotted, together with (on a separate chart) the range, that is, the difference between the highest and lowest results in the test sample (Figure 13.2). By showing both the average value and the spread of results, the state of control

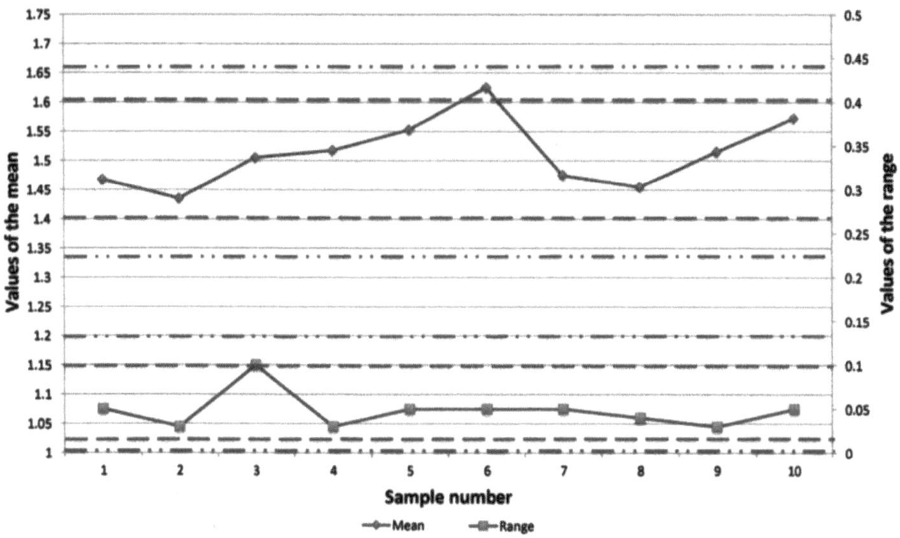

Figure 13.2: Example of a mean and range chart.

of the process is more apparent. It is common practice to put action and warning limits on these charts to help decide what is normal process variation and what variations may be indicative of potential problems. It is usual to position the warning limits at 1.96 times the standard deviation (SD) from the target value, and the action limits at 3.09 times the SD (the mean ±1.96 times the SD will cover 95% of the values, and at ±3.09 times the SD, 99% of the values, assuming that the results are normally distributed, will be covered). Alternatively, there may be upper and lower control limits on the chart, usually positioned at three SD either side of the mean.

In this example, the warning limits are shown as the dashed line, with action limits as the dotted and dashed line. It can be seen that the mean broached the warning limit on sample number 6, but the values quickly returned well into the tolerance band, and so action would probably not be necessary. However, the last three results display an upwards trend and it would be wise to note this as a possible indication that something may have changed. There are recommended rules on interpreting excursions into the warning and action limits – for example, how many samples can exceed the warning limits before action is taken, and how many results successively above or below the mean are needed to trigger investigation. The reader is referred to the many articles and textbooks on SPC, such as that already mentioned by Oakland [1], for further information. There are other control charts that can be used to monitor and control manufacturing processes, such as moving average and multiple variation charts, but in general, a simple mean and range chart will be more than adequate in most latex dipping operations.

13.4.2 Control charts for attributes

As mentioned earlier, attributes are discrete data, which can only take certain defined values. Examples would be holes in a glove or condom, or packages that are not sealed properly. Whilst run charts can be used, as with variables data, more information will be gained from a more sophisticated analysis. There are two types of attributes data: number of defectives and number of defects. The first type would be, for example, the number of products in the sample that do not comply with the specification; whilst the second could be the number of defects in a defined area of packing film, or the number of colony forming units in a microbiological plate count. In most cases, it will be the first of these two situations that will be of interest in latex dipping. Whichever method is applied, it is critical to ensure that a defect or a defective product is clearly and unambiguously defined. This should not be a problem with regulated products, where the relevant international standard will contain these definitions, but it could be a problem in other cases. It will be impossible to control a process if there is inconsistent decision-making on what constitutes a defect.

These two different types of attributes data differ, as the distribution of results from the first type, the number of defective products, is described by the Binomial distribution, whilst the distribution of the number of defects conforms to the Poisson distribution. For this reason, they are treated by different types of control charts. More details about the statistical reason for these differences will be found in Oakland's book [1] or other texts on SPC.

When the number of defectives is being used, then, if the sample size is constant, the control chart will be one called the 'np chart'. This will be the case, for example, with freedom from hole testing for condoms against ISO 4074:2014, where the sample size is dictated using ISO 2859-1:1999. Where the sample size varies, so that the proportion of defective items is studied, a different design of chart, called the 'p chart' will be more appropriate. The corresponding charts for the number of defects are the 'c chart' for constant sample sizes and the 'u chart' for varying sample sizes. As with control charts for variables, action and warning levels can be set on the charts, but very often only the upper levels are used, as a low number of defects or defectives are the desired result. However, results falling below the expected level may indicate faults with the test method or operator, or even that a process improvement has occurred. Worked examples of all of these types of control charts can be found in Oakland's book [1].

13.5 Process capability

Measures of process capability are sometimes encountered in quality management. There are three process capability indices commonly encountered: relative precision index (RPI), capability indices [process mean centred between specification limits (Cp) and process mean not centred between specification limits (Cpk)]. These indices quantify how the results relate to the required specification tolerances, and give a numerical indication of the ability of the process to yield product conforming to the specification. Note that they are only valid for variables data. Process capability for attributes data can be estimated from the average level of defective items or number of defects.

Specifications for variables data are usually expressed as the nominal value plus and minus the tolerance. The specification range is the difference between the lower specification limit (LSL) and the upper specification limit (USL).

13.5.1 Relative precision index

RPI is defined as the interval between the specification limits divided by the mean of the range 'R' of the samples measured:

$$\mathrm{RPI} = \frac{\mathrm{USL} - \mathrm{LSL}}{\mathrm{R}} \tag{13.1}$$

To assess the capability of the process, the RPI is compared to the SD. For a normal distribution, approximately 99.7% of all the results will lie within ±3 SD of the mean, so virtually all of the results will lie within a spread of 6 SD. If our experimental results show that six times the SD is approximately only equal to the tolerance spread, then the RPI is low and a slight shift in the mean could lead to an increase in rejects. A stable process that is under control will have a tolerance band that is much larger than six times the SD. In this case, shifts in the mean value of the test results can be accommodated within the tolerance band and the product will still meet the specifications. Reject material will not be produced. This does not mean however, that changes in the test results should be ignored! Any significant changes should be investigated without delay. Note that the RPI will depend on the sample size.

13.5.2 Cp and cpk indices

More commonly encountered measures of process capability are Cp, and the associated Cpk. Cp gives an estimate of what the process is capable of achieving, assuming that the process mean is in the middle of the specification limits. Cpk allows for the fact that the process mean may not be centred between the specification limits.

Cp is calculated as the tolerance band divided by six times the SD of the results:

$$Cp = \frac{USL - LSL}{6\,SD} \tag{13.2}$$

Cpk is calculated as the minimum of the difference between the USL or LSL divided by three times the SD:

$$Cpk = \frac{USL - X}{3\,SD} \tag{13.3}$$

Or

$$Cpk = \frac{X - LSL}{3\,SD}$$

whichever is the smaller

where ‘X’ is the mean.

Values greater than 1 for either Cp or Cpk show that the results fit inside the tolerance band, but for greater confidence that the production process is well under control, values of Cp and Cpk approaching 2 are preferred. As an example, consider three different sets of measurements of the length of a dipped latex product, shown in Table 13.1. The specification in this example is a length of 210 mm with a tolerance of ±10 mm.

Table 13.1: Length of a dipped latex product.

	Sample 1	Sample 2	Sample 3
1	210	210	215
2	208	210	215
3	206	212	215
4	215	210	214
5	220	209	214
6	207	213	219
7	209	210	216
8	219	208	213
9	207	207	212
10	210	210	215
Mean	211.1	209.9	214.8
SD	5.09	1.73	1.87

Inspection of these results shows that all the results lie within the tolerance band, although both sample 1 and sample 3 have individual results at or near to the limits.

The results for samples 1 and 2 are centred close to the target value, although the SD for sample 1 is much larger than that of the other samples. The process capability indices allow us to quantify these observations regarding these hypothetical results.

The tolerance band is ±10 mm, i.e., 20 mm, so the values for Cp are:

$$\begin{aligned}\text{Sample 1}: \text{Cp} &= \frac{20}{6 \times 5.09} = 0.65\\ \text{Sample 2}: \text{Cp} &= \frac{20}{6 \times 1.73} = 1.93\\ \text{Sample 3}: \text{Cp} &= \frac{20}{6 \times 1.87} = 1.78\end{aligned} \tag{13.4}$$

calculation of Cp from the data in Table 13.1.

The values for samples 2 and 3 are well above 1, showing that the value of the mean with 3 SD on either side of the mean fit easily within the tolerance band. However, the result for the Cp of sample 1 is substantially less than 1. Although the results for this particular sample lie within the specification limits, the index shows that this process is not capable, and we would expect to see a significant number of rejects over time unless effective corrective actions are applied.

Looking now at Cpk, we have a USL of 220 mm and a LSL of 200 mm. Calculation gives us these values:

$$\begin{aligned}\text{Sample 1}: \text{Cpk} &= \text{the minimum of } \frac{220 - 211.1}{3 \times 5.09} = 0.58 \text{ and } \frac{211.1 - 200}{3 \times 5.09} = 0.73\\ \text{Sample 2}: \text{Cpk} &= \text{the minimum of } \frac{220 - 209.9}{3 \times 1.73} = 1.95 \text{ and } \frac{209.9 - 200}{3 \times 1.73} = 1.91\\ \text{Sample 3}: \text{Cpk} &= \text{the minimum of } \frac{220 - 214.8}{3 \times 1.87} = 0.93 \text{ and } \frac{214.8 - 200}{3 \times 1.87} = 2.64\end{aligned} \tag{13.5}$$

calculation of Cpk from the data in Table 13.1.

The figure of Cpk = 0.58 for sample 1 confirms the result for Cp: the results show that this particular process is not in statistical control. The Cpk for sample 3 is 0.93. Thus although the value for Cp is good, the value for Cpk is less so. This tells us that the process is inherently capable, but the manufacturing process should be adjusted to centre the mean value more accurately within the tolerance band. The value of Cpk = 1.91 for sample 2 tells us that this process is statistically capable and well in control.

If excessive variation is noted within the results, then this is a warning that the process is either not under statistical control or is moving towards an out of control state.

13.6 Cumulative sum charts

Cusum charts were developed in Britain in the 1950s and are described in detail in ISO 7870-4:2011, as well as in many books on SPC. A brief description of the procedure, and the use to which it may be put is given here, but the interested reader is strongly advised to consult a statistician or specialist article before using cusum.

Table 13.2: Simulated data to demonstrate a cusum.

Sample number	Value	Sample number	Value
1	10	16	10
2	12	17	9
3	10	18	9
4	10	19	12
5	9	20	9
6	9	21	9
7	11	22	10
8	10	23	9
9	9	24	9
10	12	25	8
11	8	26	9
12	9	27	9
13	10	28	10
14	10	29	8
15	9	30	7

Unlike some control charts, cusum charts can handle both variables and attributes data, and they are particularly powerful in detecting trends and long-term changes in the property being measured.

To give an example of the power of a cusum chart, consider Table 13.2 of simulated data.

It is always difficult to get a real picture of what is happening just from a series of numbers, so let us graph them on a run chart given in Figure 13.3.

It looks as if the results may be gradually drifting down, but no real trends or shifts are apparent. Inspection of this chart may prompt a careful observer to keep an eye on the process, but would be unlikely to trigger any corrective action. However, if we show the same data as a cusum, we get Figure 13.4.

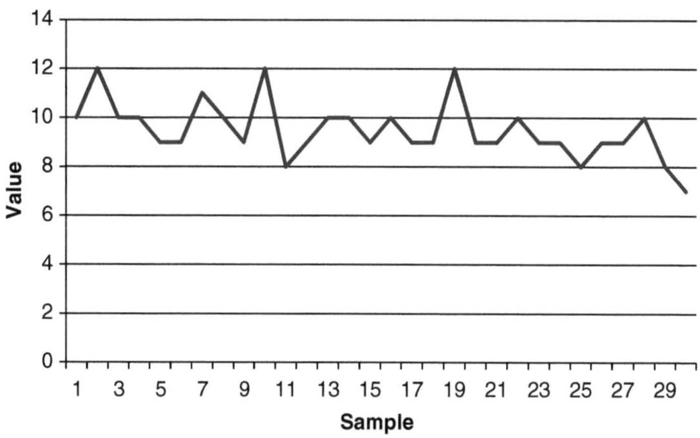

Figure 13.3: Simulated data from Table 13.2 displayed as a run chart.

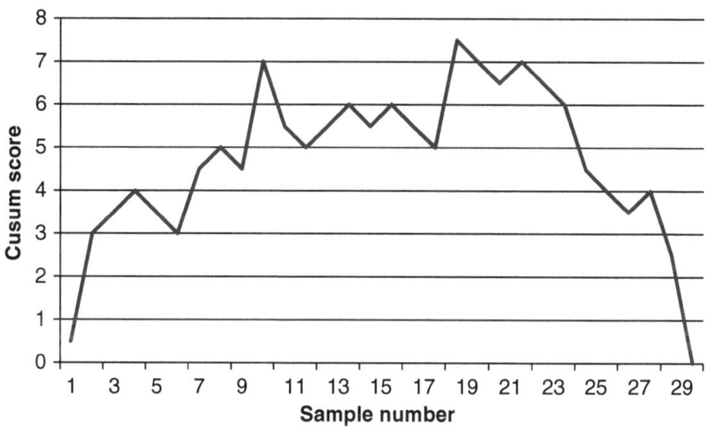

Figure 13.4: Simulated data from Table 13.2 displayed as a cusum.

A completely different picture! Up to sample number 19 the line is trending upwards, whilst from 20 onwards the trend is down. If we look at the mean of the data from the first 19 samples, we see it is 9.9; whilst the last 11 samples have a mean of 8.8. This would strongly suggest that a real change in the process occurred at this time and it should be investigated without delay. The change could be a shift change, a change in raw materials or any other factor.

So, how do we calculate a cusum? It is quite straightforward, and many mathematical and statistical data software packages will calculate cusum for you. The basic operations are these:

1. Take the benchmark or target value for the particular parameter. If you are doing it retrospectively you can calculate the average of the data.
2. Calculate the difference between each data point and the target value (or average).

3. Sum these differences.
4. Plot the cumulative sum against the run number.

Table 13.3: Calculation of a cusum, using the simulated data from Table 13.2.

Sample number	Value	Value (minus the mean)	Cumulative sum	Sample number	Value	Value (minus the mean)	Cumulative sum
1	10	0.5	0.5	16	10	0.5	6
2	12	2.5	3	17	9	−0.5	5.5
3	10	0.5	3.5	18	9	−0.5	5
4	10	0.5	4	19	12	2.5	7.5
5	9	−0.5	3.5	20	9	−0.5	7
6	9	−0.5	3	21	9	−0.5	6.5
7	11	1.5	4.5	22	10	0.5	7
8	10	0.5	5	23	9	−0.5	6.5
9	9	−0.5	4.5	24	9	−0.5	6
10	12	2.5	7	25	8	−1.5	4.5
11	8	−1.5	5.5	26	9	−0.5	4
12	9	−0.5	5	27	9	-0.5	3.5
13	10	0.5	5.5	28	10	0.5	4
14	10	0.5	6	29	8	−1.5	2.5
15	9	−0.5	5.5	30	7	−2.5	0

This can be seen in the simulated data in Table 13.3, where the mean is 9.5.

The first column is the sample number and the second column the simulated value. The third column is the difference between the value and the target value or mean – in this case the mean. Finally, the fourth column is the cusum score – the sum of the differences between the value and the benchmark in column 3 (columns 5–8 just continue to list the data). So it can be seen that if the values are consistently above the benchmark the cusum score increases, whilst if they are consistently below the benchmark the cusum will decrease. A persistent change in the values will therefore result in a change of slope.

It must be emphasised that the cusum technique *must* be used with care, and preferably by people trained and experienced in statistical data analysis. A cusum of real data will show many small changes in slope, caused by normal variation. The parameters of the chart must be selected carefully, to ensure that the slope, and any changes in the slope, are of the correct magnitude for the required sensitivity. We need to be sure that the change of slope is a real effect and not just random variation. Various statistical decision procedures are used to confirm statistical significance, which will normally be included in any statistical software package. There is not the space available in this chapter to deal with these decision procedures in detail – the interested reader is referred to the Standards or one of the many excellent texts on the subject to know more.

Used wisely, and preferably in conjunction with the standard Shewhart process control charts, cusum charts can yield valuable information about a process, and in particular, about sustained changes that may not otherwise be easily detected.

They are thus extremely useful in both process control and troubleshooting and, correctly used, can direct the investigator to exactly when a change in a process occurred.

13.7 Pareto analysis

Pareto analysis, also known as the 80/20 rule, is another way of saying that most of the problems arise from a few causes. The 80/20 comes from the empirical estimation that approximately 80% of problems can be traced to approximately 20% of causes. Data can be arranged according to priority – this could be number of defects, cost, importance to the customer and so on. Once arranged in this way, corrective action can be targeted to address the important problems.

Detailed information about Pareto analyses is available in the general quality control literature, so only a brief summary of the procedure will be given here:

- The first step is to make a list of all of the defects, defined as accurately as possible.
- Having listed all of the elements, the next step is to assign a parameter to them.
- The same parameter must be used for all, to allow comparison, and could be frequency of occurrence, cost, significance to customer and so on.
- Once the parameters have been listed, the different components in the analysis can be ranked in order from the greatest (also known as the ‘vital few’) to the least (the ‘trivial many’).
- The cumulative frequency of the different components can now be calculated as a percentage and the Pareto curve drawn.
- The chart will now show which of the various elements have been identified as the most important, depending on the criteria applied, and resources can be targeted to address them.

As an example, consider this following analysis of glove defects. The defects have been categorised as to cause and tabulated in Table 13.4. They have been ranked according to frequency and a cumulative frequency calculated in

Table 13.4: Table of glove defects and their frequency.

Defect	Number	Percentage
Bead defects	133	38.3
Body/finger holes	13	3.8
Crotch holes	98	28.2
Cupping	5	1.4
Dirt	13	3.8
Foreign body	11	3.2
Latex runs	40	11.5
Other	22	6.3
Stripping creases	12	3.5
Total	347	100.0

Table 13.5: Cumulative frequency of the glove defects in Table 13.4.

Defect	Number	Percentage	Cumulative (%)
Bead defects	133	38.3	38.3
Crotch holes	98	28.2	66.5
Latex runs	40	11.5	78.0
Other	22	6.3	84.3
Body/finger holes	13	3.8	88.1
Dirt	13	3.8	91.9
Stripping creases	12	3.5	95.4
Foreign body	11	3.2	98.6
Cupping	5	1.4	100
Total	347	100	–

Table 13.5. From this, a Pareto curve has been drawn in Figure 13.5, showing the percentage of defects, ranked according to frequency, together with the cumulative frequency.

This highlights the fact that bead defects, crotch holes and latex runs account for 78% of the defects, and any quality improvement programme could prioritise these. However, although stripping creases only account for 3.5% of the observed defects, it is possible that they could be easily removed, for example, by increasing the maturation period of the latex, increasing vulcanisation temperatures slightly or adding more powder to the coagulant. Despite the relatively low frequency of stripping creases as a defect, the company could decide to fix these early, as well as the more frequent defects.

It should be noted that Pareto analysis is basically just a method for allowing a situation where there are multiple defects to be clarified. Unlike some of the

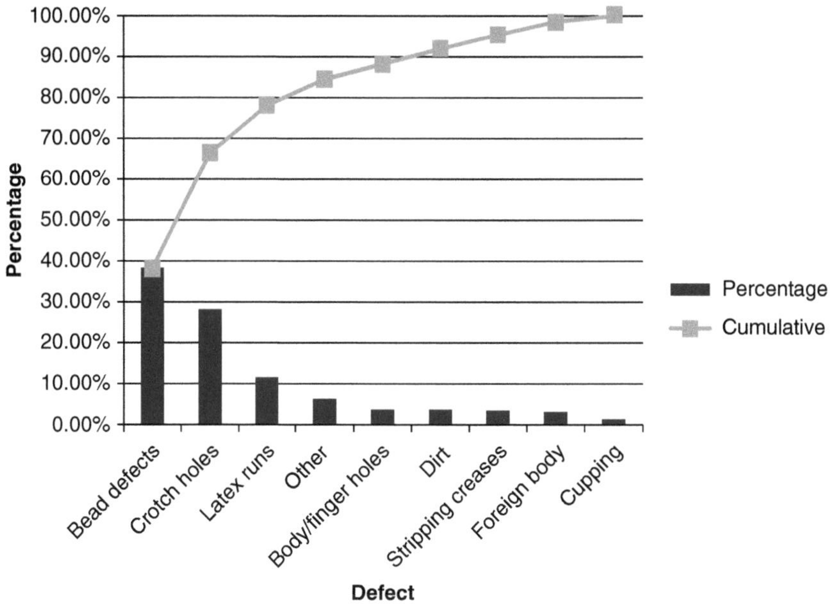

Figure 13.5: The Pareto curve from the data of glove defects given in Tables 13.4 and 13.5.

other techniques described in this chapter there is no underlying mathematical nor statistical basis. Nevertheless, it provides a simple method of cutting through what may be a confusing quality situation and can help limited resources to be effectively employed to improve quality. It is only a guideline and not to be followed too rigorously: if some of the 'trivial few' can be fixed quickly and easily, then do it.

References

1. J.S. Oakland in *Statistical Process Control*, 6th Edition, Routledge, Oxford, UK, 2008.
2. ISO 7870-1:2014: Control charts – Part 1: General guidelines.
3. ISO 7870-2:2013: Control charts – Part 2: Shewhart control charts.
4. ISO 7870-3: 2012: Control charts – Part 3: Acceptance control charts.
5. ISO 7878-4:2011: Control charts – Part 4: Cumulative sum charts.
6. ISO 7870-5:2014: Control charts – Part 5: Specialized control charts.
7. ISO 7870-6:2016: Control charts – Part 6: EWMA control charts.

14 Process validation

14.1 Introduction

Many dipped latex products such as condoms, catheters and surgical gloves are classified as medical devices, and have to comply with regulatory requirements. As such, the manufacturing processes have to be validated. It is a requirement of Quality Management Systems, such as International Organization for Standardization (ISO) 9001:2015 (section 8.5.1) and also ISO 13485:2016 (section 7.5.6), that *'The organization shall validate any processes for production and service provision where the resulting output cannot be verified by subsequent monitoring or measurement... Validation shall demonstrate the ability of these processes to achieve planned results consistently.'*

It is also required by European standard, BS EN 46001:1997 (European quality system for medical devices) and various other good manufacturing practice guidelines. There are several good guides for the process validation of medical devices, and this chapter will simply provide an outline of what the latex dipping factory should do in order to ensure that their processes are properly validated. Process validation is more than just a necessary regulation however. Properly and sensibly carried out, process validation will lead to a much better understanding of the process, will enable a better response to quality problems, more accurate troubleshooting and the development of a process which is optimised for cost and quality.

Note also that any products and services from external suppliers also fall under the validation requirements of ISO 9001:2015 and ISO 13485:2016.

Process validation, as defined by the Global Harmonization Task Force '*Quality Management Systems – Process Validation Guidance*' [1] is '*establishing by objective evidence that a process consistently produces a result or product meeting its predetermined requirements*'. In other words, a proper process validation will generate results that give confidence that the process will deliver the required results all the time, every time.

As noted in the quote from ISO 9001:2015 and ISO 13485:2016 above, there are occasions when verification may be considered as an alternative to validation. Verification can be thought of as '*making the thing right*', whilst validation may be considered as '*making the right thing*'. In other words, verification ensures that the manufacturing process has been followed correctly: validation ensures that the product has the correct characteristics for safe and effective use. Process verification would therefore entail the continuous monitoring of those parameters defined as 'critical to quality' (CTQ). The problem is that there can be many CTQ parameters in latex dipping, from raw and compounded latex factors, through dispersion quality to the settings on the production plant. In addition, many of the relevant tests are destructive and 100% inspection does not guarantee 100% quality. It is well accepted that quality cannot be inspected into a product. Finally, validation is more likely to

https://doi.org/10.1515/9783110638097-014

capture the effect of unknown or incompletely understood variables. For these reasons, most regulatory bodies will demand that the process is validated.

Process validation can be considered to be comprised of four steps:
- Design qualification (DQ).
- Installation qualification (IQ).
- Operational qualification (OQ).
- Process qualification (PQ).

Note that, once validated, there will need to be a programme of revalidation of the process at appropriate intervals, as described below in Section 14.7.

It is recommended that any process validations are managed by a dedicated validation team, and carried out as part of a properly constituted master validation schedule. It is best if the team is multidisciplinary, covering production and engineering, as well as quality control and quality assurance. The team need not be fixed, with specialists drafted into the team as required. The team will identify and define the requirements of the validation in as much detail as possible through a validation plan. The elements of process validation are considered in more detail in Sections 14.2–14.5.

14.2 Design qualification

DQ will demonstrate (using objective evidence) that the design complies with the appropriate technical specification and will identify any regulatory implications of the equipment and process. In practice, it is typically a paper exercise, reviewing and approving drawings and specifications and so on. Where the DQ covers equipment, it will usually be carried out by the machinery supplier.

14.3 Installation qualification

IQ will apply to any new, modified or refurbished equipment. It is carried out to demonstrate that the equipment has been installed or modified correctly, it works as it should and that the positioning, utility supply, calibration, and so on., are as required. An IQ will typically consist of a list of requirements that will be checked against performance and ticked off and reported when complete.

14.4 Operational qualification

In most of the process validations that will be encountered, OQ is where the bulk of the work will be carried out. It will define how the process operates best, identify the operating limits and critical parameters, what operator training will be required, which raw materials are needed, what calibration and cleaning schedules are necessary and so on. OQ will probably require a series of trials carried out to clearly defined conditions, to determine the maximum and minimum operating conditions that will produce satisfactory product (the so-called 'worst case' conditions). The outputs from OQ will include process capability measures (Cp and Cpk, as described in Chapter 13) and optimum operating conditions. In short, OQ will show how the process is to be run to give the required products at the optimum yield and minimum cost. In many cases, previous experience with similar equipment and processes can mean that a lot of the OQ parameters will already be known to some degree, but no matter how much is already known, any new or modified machine or process *must* be validated. The current versions (at the time of writing) of both ISO 9001:2015 and ISO 13485:2016 are moving towards a risk-based approach, and it is strongly recommended that a risk analysis is incorporated into OQ, to answer questions such as 'what can go wrong?', 'what are the consequences if anything goes wrong?', 'how do we detect it?', 'how do we prevent it?' and so on.

14.5 Process qualification

The final stage in a process validation is PQ, which should show that the process, as defined in the OQ stage, is stable and repeatable. The process capabilities will be confirmed and the specifications for the product and process should also be confirmed. Typically, this will be carried out by preparing several (at least three) consecutive batches to the process as defined in OQ, and establishing that they meet all the required specifications. If the OQ stage has been carried out properly, PQ will be a relatively short and simple process.

14.6 Reporting

All of these stages should be described in properly written protocols and reported in a formal signed-off report. Both the protocols and reports can be in a pro forma layout, but all of the data must support the requirements of Section 14.5. The conclusions must be accessible in the event of an audit. Validation documentation will come under the document control procedures in ISO 9001:2015 and ISO 13485:2016. The protocol should show what to measure, how to measure it, how many to measure and when to measure. The accept and reject criteria must be clearly defined in advance, as will the way in which the results will be reported. A process validation need not generate an enormous amount of paperwork – just the right paperwork.

14.7 Revalidation

One thing not to be forgotten in process validation is the need for revalidation. Many companies operate some form of continual improvement programme, and these programmes can identify and implement changes to the process. In addition, ISO 13485:2016 includes the requirement for a manufacturer to maintain the suitability, adequacy and efficacy of any medical device they produce, and also to revalidate [section 7.54.6 (f)]. Similarly, any significant change in the product or raw materials should trigger a revalidation, and any such significant change in the process must be revalidated. Any negative trends in quality are also a sign that things may have changed and a revalidation is in order. An obvious question here is 'what is a significant change?' Each situation must be assessed on its merits by the validation team, but one pointer is any change that may impact the quality of the product. It is strongly recommended that the justification for any decision regarding what constitutes a 'significant change', and whether to revalidate or not, is recorded in writing and filed. Finally, even without any major changes, there will be the need for a periodic revalidation. There will generally be small changes and improvements made to a process over the years, none of which is large enough by itself to require a revalidation, but taken together they can add up to a major change. The frequency of routine revalidation will depend upon the process and product, but will typically be every 3–5 years. Such a revalidation will typically consist of taking and quarantining three production batches and testing them to a higher level than normal. If the batches comply with the requirements, then the process can be considered revalidated and reported as such.

14.8 Concurrent and retrospective validation

The type of process validation described above is called 'prospective validation' and should be considered as the norm, to be used whenever validation is required. However, there may be cases where it is necessary to validate an ongoing process that has not been validated. Examples could be a change in the market for a product, the original validation has been found to be flawed or changes in the regulations have taken place. In this case, alternative validation processes may possibly be applied. These are known as 'concurrent' validation, and 'retrospective' or 'historic' validation.

14.8.1 Concurrent validation

Concurrent validation uses data from an ongoing process to carry out the validation. In this case, the rationale for concurrent validation must be documented. In addition, the process must be in a state of statistical control – it is not possible to validate a process where the results of the process vary in an uncontrolled manner. To carry out a concurrent validation, the critical parameters of the production must be monitored and recorded. Several batches of product from the process should be quarantined and tested to the requirements specified in the protocol. It is recommended that the testing should be carried out to a higher acceptance quality limit, preferably using a larger sample size. If the product passes the requirements, it can be released and the process considered validated. As with prospective validation, a fully signed-off protocol and report are needed.

14.8.2 Retrospective validation

An alternative to concurrent validation could also be the use of historic data – 'historic' or 'retrospective' validation. For retrospective validation to be acceptable, the process must not have changed between the time the data were recorded and the time of the validation. In addition, the process must have been under good statistical control when the data used for the validation were gathered. As with other validation processes, the rationale, protocol, results, reports and so on, must be documented.

14.9 Example of a process validation

An example of a simple process validation could be that of a condom foiling machine. A condom foiling machine takes the rolled condom and carries it to sealing jaws, where the correct quantity of lubricant is injected as the condom is sandwiched between two strips of a laminated foil. The foil is then sealed by applying heat and pressure to the edges. Printing of the foil with the required details (batch number, expiry date, manufacturer and so on) may be carried out during the foiling process or in a separate off-line operation.

Critical parameters could be defined as machine speed, sealing jaw temperature and pressure, lubricant quantity, and print quality and legibility. Hence, the required outputs from the validation would be seal integrity, absence of any damage to the condom, correct location of the condom in the foil, correct printing and the correct amount of lubricant. Condom foiling machines can be bought from several suppliers and we can assume that DQ has been carried out by the supplier: it is therefore necessary to carry out IQ, OQ and PQ.

The validation team should be gathered and initially could include members from the engineering and production to the quality assurance departments, although others could be included if required.

14.9.1 Installation qualification

The following work should be carried out (and note that this list is not exhaustive):
- Is there a correct signed-off protocol?
- Is the location of the machine satisfactory?
- Is the supply of electricity, compressed air and any other services satisfactory?
- Do the sealing jaws heat up properly and adequately quickly?
- Is the temperature control and indication correct?
- Is the line speed and control of the line speed satisfactory?
- Is the feed of the upper and lower webs of the foil satisfactory?
- If relevant, is the printer working properly?
- Does the lubricant pump work properly, does it deliver the correct amount of lubricant and can it be adjusted satisfactorily?
- Do the user controls work satisfactorily?
- Are all relevant health and safety controls working correctly?

When all of this is done, a report should be written and signed-off. The report can be a tick-list recording the different operations, but all raw data should be recorded.

14.9.2 Operational qualification

Once IQ has been completed, the validation can move on to OQ. At this point the validation team may include members from the marketing, regulatory, and research and development departments. As mentioned earlier, the purpose of OQ is to define the optimum operating parameters and the range of parameters that will yield acceptable product. It will also quantify process stability and capability, define any calibration requirements, and identify any raw material and training requirements. In this case, OQ could include the following areas, which will, of course, be fully described in a signed-off protocol:
- Run the line speed at 90, 95, 100, 105 and 110% of the target production speed. This will help to define operating speed limits that can be used, which may be useful if it is necessary to change the line speed to solve sealing problems or cope with varying demands on production.
- Use sealing jaw temperatures of 90, 100 and 110% of the target production temperature. As noted above, this information could be of use if the production conditions need to be varied.
- Maintain the sealing jaw pressure at the supplier's recommended pressure. If this gives acceptable results, it is not normally necessary to vary the pressure.
- Run the lubricant dosage at the internal specification for the condom type. If more than one condom type, each with a different lubricant level, is to be run on the machine, then these will also need to be covered.

- The machine will need to be allowed to stabilise at each of these conditions, following the manufacturer's recommendations.
- Run the machine for a period of time at each of the experimental conditions, e.g., 1 h.
- Test the product from each of the experimental conditions according to the protocol. These tests could include seal integrity, freedom from holes in the condom, the correct location of the condom in the foil, correct printing and the correct amount of lubricant.
- Finally, analyse the results and write the report. The report will typically include the acceptability of the product under each of the experimental conditions, Cp and Cpk, yield data, any problems encountered, any potential failure modes encountered and any training requirements identified. It is always a good idea to carry out a risk analysis during OQ.

14.9.3 Process qualification

Once the optimum operating conditions have been established in the OQ stage, PQ can be carried out, usually by running multiple batches under the conditions defined in the OQ stage, and then testing the sealed condoms to the specification requirements.

This example covers a very simple validation, which could be completed within a few days. More complex validations can take longer, especially in the OQ stage. For this reason, it is important that the validation team consider the design of the trials very carefully, ensuring that all critical parameters are covered fully, but time is not wasted on experiments that do not add to the knowledge required for validation. It can be very easy to become overloaded with data, which may obscure important results. Time spent in planning will save a lot of time, energy and costs later in the validation. Finally, remember that a validation can reveal a lot about the process that could be of value in other areas, such as troubleshooting and process improvement. Don't lock the results away in a file and forget about them!

Reference

1. *Quality Management Systems – Process Validation Guidance*, 2nd Edition, The Global Harmonization Task Force, 2004. https://www.fda.gov/OHRMS/DOCKETS/98fr/04d-0001-bkg0001-10-sg3_n99-10_edition2.pdf.

15 Some analytical methods used in latex dipping operations

Many dipping companies will have the facilities to conduct tests on the incoming raw latex and these are adequately described by the relevant international standard. The more common International Organization for Standardization (ISO) standards are listed in Table 15.1 (versions given at the time of writing).

Table 15.1: List of common ISO standards for latex.

Test	ISO standard number
Sampling of rubber latex	ISO 123:2001
Mechanical stability time (NR latex)	ISO 35:2004
Total solids content	ISO 124:2014
Dry rubber content	ISO 126:2005
Volatile fatty acids (NR latex)	ISO 506:1992
Alkalinity	ISO 125:2011
Potassium hydroxide number (NR latex)	ISO 127:2012
pH	ISO 976:2013
Coagulum content	ISO 706:2004
NR latex specifications	ISO 2004:2010

NR: Natural rubber

Most other chemical analyses on latex or dipped latex products are outside of the scope of all but the largest latex dipping companies. Further information for those interested can be found in the excellent book by Loadman [1]. However, there are some simple and useful tests that can be carried out with limited equipment.

Test strips are available (e.g., 'Merckoquant' or 'Quantofix') which can estimate the concentration of various materials in water by simply dipping the strip into the liquid and then comparing any colour change on the test strip against a scale. Their accuracy is reasonable and they can be very useful in troubleshooting, e.g., by checking the level of contaminants such as copper or iron in the latex or process water. Test strips are also available to measure total water hardness. A full list of the test strips available can be obtained from the manufacturers' websites.

A very simple qualitative method for determining if there is any residual dithiocarbamate in a rubber article is to moisten a copper coin and press it firmly against the rubber for a minute or so. Any dithiocarbamate will react with the copper in the coin and give a brown stain of copper dithiocarbamate. The darker the stain, the more dithiocarbamate there is in the rubber.

A more accurate estimation of residual dithiocarbamate accelerator in a dipped latex product can be made easily if there is access to an ultraviolet/visible

https://doi.org/10.1515/9783110638097-015

spectrophotometer. A known weight of the product is extracted with a known volume of a solvent such as dichloromethane. Any residual dithiocarbamate accelerator will be extracted into the solvent. The dichloromethane extract is then shaken with a solution of copper sulfate in water. Any dithiocarbamate accelerator extracted into the dichloromethane layer will react with the copper to give the brown copper dithiocarbamate, which will remain in solution in the dichloromethane layer. The intensity of the brown colouration is concentration dependent, and the amount of dithiocarbamate in the solution can be estimated by measuring the absorption at a fixed wavelength (e.g., 450 nm) and comparing the absorbance against a previously prepared calibration curve.

A larger chemical laboratory may also have the capability of measuring extractable latex proteins in NR. The usual method is the modified Lowry method, described in American Society for Testing and Materials, ASTM D5712 and ISO 12243:2003. The Lowry assay is a colorimetric method, based on the complexing of copper with protein. The copper complex is reacted with Folin reagent, and the intensity of the resulting blue colour is proportional to the concentration of proteins in the sample and can be estimated spectrophometrically. The original Lowry method was very susceptible to interference by other materials such as surfactants and buffers, but various modified methods have been developed to reduce, but not eliminate, these interferences.

In this method, the product is shaken with an aqueous buffer solution of pH 7.4. After filtering, the extract is treated to precipitate the dissolved proteins. The proteins are separated by centrifugation, redissolved and quantified in a colorimetric reaction against a standard.

Although the current method is much less susceptible to interferences from external materials, bear in mind that materials such as condom lubricants or glove finishing treatments could affect the results, and suitable precautions, such as appropriate control samples, should be employed.

Reference

1. M. John and R. Loadman in *Analysis of Rubber and Rubber-like Polymers*, Springer Science, Berlin, Germany, 1998.

16 Allergies, proteins and *N*-nitrosamines

Most of those who work in the latex dipping industry will be aware of the problems that can exist with irritations, allergies, extractable latex proteins and *N*-nitrosamines. Some of the chemicals used in the latex dipping industry are toxic, to one degree or another, and in certain circumstances their use can give rise to problems. These problems include irritation (also called allergic contact dermatitis) and other allergic reactions, both immediate and delayed.

16.1 Allergic reactions

The allergic reactions encountered with rubber products are 'type I' and 'type IV' reactions. The type I reaction is usually immediate although it may be delayed for a few hours after contact. The type IV reaction can be delayed for several days after exposure. In general, an individual needs to be sensitised – usually by previous contact with an allergen – to suffer a reaction.

Many materials can cause a reaction in sensitised individuals, not just rubber chemicals. However, rubber accelerators are amongst those possible irritants commonly met in day-to-day life. The allergic reaction is usually local, although in severe cases it can be systemic. The commonly occurring symptom is a rash over the affected area, which may be painful or itchy. It is recommended that manufacturers try to minimise the concentration of residual accelerator and other potential irritants in dipped rubber articles, especially those which may be in long-term or repeated contact with the user, such as gloves, or those which contact mucous membranes, such as condoms. This can be achieved through formulation, compounding procedures, and efficient leaching and washing of the articles during production. Note however, that residual dithiocarbamate accelerators also have a protective effect against degradation, and if these accelerators are present it may be necessary to balance the risk of allergy against degradation. There are clinical tests available, for example, those described in International Organization for Standardization standard, ISO 10993 – Part 10 – Tests for irritation and skin sensitisation, which allow any risks to be quantified.

16.2 Extractable latex proteins

The problem with allergic reactions to proteins that can be extracted from natural rubber (NR) is one of the most, if not the most important problem the NR-latex industry has faced over the last few decades. This topic has been exhaustively debated in the scientific literature, regulatory meetings and conferences over the

https://doi.org/10.1515/9783110638097-016

past few decades, and this section will merely try to provide a brief review of the current situation and where it leaves the latex industry at present.

The problem with extractable latex proteins is generally considered to have surfaced with a vengeance in the late 1980s, and is presumed to be linked with the sudden surge in demand for surgical and examination gloves as the impact of human immunodeficiency virus (HIV)/acquired immune deficiency syndrome (AIDS) became apparent. It is thought that this sudden expansion of the market resulted in a flood of improperly manufactured NR gloves being produced, without sufficient leaching, and hence containing more extractable materials than was previously the case. The concerns over HIV/AIDS would also result in the gloves being used by and on people who may not have previously been exposed to rubber so frequently – healthcare workers, patients and the general public.

Proteins are all around us and are an indispensable part of life. The fundamental building blocks of proteins are amino acids, and proteins are a sequence of these amino atoms polymerised in a specific conformation. Although there are a limited number of amino acids found in proteins, the many ways that they may be sequenced into the protein molecule means that there are very many different types of proteins found in nature. A few of these, including some found in NR latex, have the potential to provoke an allergic reaction in certain individuals. In these cases, the body's immune system wrongly identifies a protein as a hostile invader and activates defence mechanisms, including the production of histamines, which in turn triggers an inflammatory response in the body. The consequences of this set of circumstances can range from a mild reaction to potentially fatal anaphylactic shock. If a person is atopic – that is, predisposed to an allergy, or sensitised, possibly through prior exposure to allergenic proteins – then a minute amount of protein can precipitate this response. Sensitisation can occur through exposure to proteins in food (peanuts, shellfish and kiwi fruit are notorious for this) or the environment, and not necessarily through NR latex.

The level of proteins in NR latex will vary to some extent, but will typically be around 1 to 2%, the bulk of which are in the aqueous phase. Over 200 different polypeptides have been identified in fresh NR latex, but only some have any allergenic potential [1]. The bulk of the proteins will be removed when the latex is concentrated by centrifuging (but not by evaporation). Of those water-soluble proteins remaining, which may be incorporated into the product, many can be removed if adequate and efficient washing is incorporated into the production processes. Of those that are left, only a few specific proteins can cause an allergic reaction, and those thought to be responsible are the subject of intensive study.

To reduce the risk it is necessary to ensure that the washing operations carried out during the manufacturing process are thorough and efficient. Latex allergy is a problem caused by extractable proteins: if more of the proteins are extracted during manufacture, the risk is correspondingly reduced [2, 3].

The proteins in NR latex are not only present in the aqueous phase, but are also strongly bound to the surface of the latex particle. Although associated with the surface, these proteins may also trigger an allergic reaction. There are ways in which these bound proteins, as well as those in the aqueous phase, can be greatly reduced, by, for example, treatment with special enzymes or complexing with other chemicals. As a result there are several 'deproteinised' NR latices available from suppliers, some of which are now being successfully used to dip products with very low extractable protein levels. Other types of *Hevea* latex, for example, from the guayule plant, containing different proteins to those found in NR, are also thought not to be allergenic, although that could be because their limited use to date has meant that exposure to these proteins has been restricted. Developing these alternative sources of latex is an active industry today.

However, in many cases the only viable solution to avoid extractable latex proteins is a synthetic alternative, such as acrylonitrile-butadiene or polychloroprene latex, or artificial latices such as synthetic *cis*-1,4-polyisoprene latex.

As mentioned earlier, the usual route of exposure to extractable latex proteins is through skin contact, especially in the presence of moisture. However, if the product is powdered with cornstarch – a common powdering material, especially for surgical gloves – it has been shown that latex proteins can be adsorbed onto the starch, which then offers an exposure route to the proteins through inhalation of the powder. The amount of protein adsorbed onto the powder does not appear to be linked directly to the amount of extractable proteins in the glove; rather it is thought that other factors, such as manufacturing conditions and powder type may be involved [4]. For this reason, powder-free gloves are widely used, e.g., in hospitals where patients' immune response may be compromised.

One question which is frequently asked is 'what is a safe level of extractable latex protein?' The answer is 'nobody knows'. A sensitised person can react to very low levels of allergen, although in general, the lower the level, the lower the risk. Work has shown that gloves with less than 100 μg/g of extractable protein have a low allergenic potential [2, 3], and this level of extractable protein on dipped NR articles can be attained by careful manufacturing and thorough washing of the products.

16.3 *N*-nitrosamines

N-nitrosamines are chemical compounds, some are which are known to cause cancer in animals, and are believed to be carcinogenic to humans. They can be found in the environment, in many foodstuffs and cosmetics, although well below the levels believed to be harmful. Of particular relevance to the latex industry, however, is that these compounds can be formed in dipped rubber articles through the reactions of many of the accelerators used in the industry. In 1982, two German workers,

Spiegelhalder and Preussman, published results showing *N*-nitrosamines in a range of rubber articles, and in the same year the *Bundesgesundheitsamt* (the German federal health agency) proposed upper limits for *N*-nitrosamines and *N*-nitrosatable amines (that is, materials that can be converted to *N*-nitrosamines) in baby teats and soothers. Regulation of these materials was closely followed by the US Food and Drug Administration (FDA) and regulatory bodies in several other countries.

N-nitrosamines are compounds with the general structure shown in Figure 16.1, where 'R' is an alkyl or aryl group, and they are formed by the reaction between secondary amines and nitrous acid (Equation 16.1).

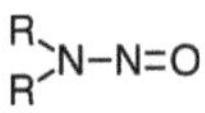

Figure 16.1: General structure of *N*-nitrosamines.

$$R_2NH + HNO_2 \rightarrow R_2N\text{–}NO + H_2O \tag{16.1}$$

Secondary amines are formed by the decomposition of many rubber accelerators, and thus can be present in vulcanised rubber articles. These secondary amines may be nitrosated by nitrous acid in the atmosphere or, if ingested, in the stomach, to form the corresponding *N*-nitrosamine. This possible exposure route is the driving force for much of the relevant legislation. Exposure to skin or mucous membranes, such as will occur with gloves or condoms, is thought to present much less risk. If the vulcanising or drying ovens on a manufacturing plant are heated by open gas flames, the nitrous oxide which may be produced during the gas combustion can provide a route for the formation of *N*-nitrosamines. If direct gas-fired heating is employed on the dipping line or drying ovens, then monitoring of the products for nitrosamines is recommended.

There is some variation of the concentrations of *N*-nitrosamines and nitrosatable amines allowed in the various standards around the world. For example, in North America, the FDA allows a total of 60 ppb of *N*-nitrosamines in bottle teats and soothers, but no single nitrosamine can be higher than 10 ppb. In Europe, the European Directive 93/11/EEC states that migration into the artificial saliva test medium must not exceed 10 µg/kg (10 ppb) for *N*-nitrosamines and 100 µg/kg for *N*-nitrosatable amines. Note that there is a difference in extraction methods. In Europe, an artificial saliva is used to give some simulation of the amount of any materials that could enter the digestive tract. In North America, a dichloromethane extraction is used, which is more likely to extract the total amount and thus give a higher figure. It is not possible therefore to make a direct comparison between the two. The maximum allowable concentration of *N*-nitrosamines in balloons and toys is laid out in the European toy safety standard, BS EN 71-12:2013, at 0.05 mg/kg (50 ppb) for *N*-nitrosamines and 1 mg/kg for nitrosatable amines. Condoms and materials used in food-contact applications are similarly subject to statutory or advisory limits in some parts of the world, notably Germany. There are also German guidelines (TRGS 552) for the concentration of airborne *N*-nitrosamines in the workplace, currently set at 1 μg/m^3.

Analysis of *N*-nitrosamines is usually by liquid or gas chromatography. Owing to the very low concentrations that need to be measured, a meticulous analytical process is required in order to avoid interference, cross-contamination or sample losses. In addition, a very sensitive method must be used. Although several detection processes are possible, most methods use thermal energy analysis, which is both sensitive and specific for *N*-nitrosamines. As mentioned above, any *N*-nitrosamines and nitrosatable amines are extracted from the product either by an aqueous artificial saliva mixture or a solvent such as dichloromethane or methanol. If the aqueous route is used, the solution is then extracted with dichloromethane in the work-up procedure, prior to quantification.

As shown earlier, the presence of secondary amines is a requirement for nitrosamine formation. These secondary amines generally come from the vulcanisation accelerators used in latex formulations, particularly dithiocarbamates and thiuram sulfides. In many cases NR-latex producers may add a small amount of thiuram disulfide to NR latex as a secondary preservative.

For manufacturers selling into these markets the raw latex, latex formulation and manufacturing processes must be such that the final levels of *N*-nitrosamines and *N*-nitrosatable amines are as low as possible. As described in Chapter 1, there are accelerators that do not give secondary amines, for example, xanthates or xanthogen polysulfides. Alternatively, dithiocarbamate accelerators are available based on high molecular weight (MW) alkyl groups, e.g., dibenzyl and diisononyl dithiocarbamates. These give residual amines of a high MW that are much less soluble (and therefore less likely to be extracted in use or analysis), and are believed to be less carcinogenic than their lower MW variants. It must be noted that these accelerators are very different in behaviour to the more usually used zinc diethyldithiocarbamate or zinc di-*n*-butyldithiocarbamate, and as such will require modification (sometimes significant modification) to the manufacturing process. However, it is perfectly possible to make good-quality products with low or zero levels of *N*-nitrosamines and *N*-nitrosatable amines. As usual, the technical departments of the chemical suppliers will be able to help and advise on such matters.

References

1. T. Palosuo, *Revue Française d'Allergologie*, 1997, **37**, 8, 1184.
2. E. Yip, K. Turjanmaa, K.P. Ng and K.L. Mok in *Proceedings of the Latex Protein Allergy: The Latest Position International Conference*, Paris, France, 1995.
3. E. Yip, T. Palosuo, H. Alenius and K. Turjanmaa, *Journal of Natural Rubber Research*, 1997, **12**, 2, 120.
4. V.J. Tomazic-Jezic, A.D. Lucas and B.A. Sanchez, *Journal of Immunoassay and Immunochemistry*, 2004, **25**, 2, 109.

17 Potential future trends in the latex dipping industry

17.1 Introduction

The production of articles by latex dipping first started in the early 20th century, following the discovery that natural rubber (NR) latex could be concentrated by centrifugation, and was patented by W.L. Utermark in 1914. Prevulcanisation of NR latex, discovered by P. Schidrowitz in 1920, meant that useable rubber articles such as condoms and gloves could now be made directly from latex. Previously, all dipping was from a solution of rubber in a flammable and dangerous solvent such as xylene. So, a century later, what has changed? There are now many more chemicals available with which to compound latices, and dipped products nowadays can have far better strength, resilience and ageing resistance. But what could the future bring? Here are some personal thoughts.

17.2 Possible manufacturing processes

17.2.1 Spray coating

Dipping is an easy way to make complex shapes, but it does suffer from some disadvantages. Straight dipping on chain plants requires long, relatively narrow latex dip tanks, and currently it is very difficult, if not impossible, to achieve good agitation of a latex in these tanks. The very process of dipping means that there will be flow, of latex, coagulant or both, and this will mean that there will be some thickness variations along the length of the product. In addition, dipping lines, especially chain dipping plants, are big. Some glove plants can be 100 m long and maybe 10 m or more high. Many of these factors could be overcome by spraying the latex onto the former. Although latices are shear sensitive, it is possible, with the correct equipment and formulation, to spray latex onto a former. One of the problems however, is to ensure that complex shapes, such as glove formers, are adequately covered. Simpler shapes, such as condom formers or baby teats, would be much easier to spray coat.

Electrostatic spraying, where the sprayed particles carry a charge and will be attracted to a former that is earthed, is sometimes used to apply a flock coating to gloves. However, it is not known if such a process would work with latices.

https://doi.org/10.1515/9783110638097-017

17.2.2 Electrodeposition

Electrodeposition of latices has been covered in Chapter 8. Whilst it is possible to form a rubber film from a latex in this way, the electrolysis of the water brought about by the electrical current means that experiments to date have shown that only a poor-quality film is obtained. Whilst it is difficult to see how this could be overcome, the control of film thickness and the possibility of creating hole-free films – it could be expected that the current, and latex particles, would preferentially flow to areas where the developing film was weakest – are attractive attributes, and could be worthy of further study.

17.2.3 Drying and vulcanisation

Drying of the latex film is usually accomplished by heat, either radiant heat, hot air or a mixture of both. Limited experiments using a dark-coloured former to increase the absorbance of the dipped film did not show any significant improvement in the rate of drying, although it is possible that using an infrared wavelength tuned preferentially to the absorbance of water, rubber or the former could be advantageous.

Microwave radiation is often touted as a possible technique to be used for heating in latex dipping. Any objects with an electrical dipole will heat up rapidly when exposed to microwave radiation. Unfortunately NR is non-polar, and therefore has a poor response to microwaves, although polychloroprene (CR) and nitrile-butadiene rubbers will heat up more readily. Microwaves will rapidly heat up water, although this effect will, of course, diminish as the water is lost. Some ceramics (e.g., silicon carbide) will heat up rapidly when irradiated by microwaves and formers made from such materials can be used in latex dipping. Preferentially heating the former in this way will help the film dry 'from the inside out', reducing skinning, and therefore facilitate the removal of water from the film. The use of microwave heating on a continuous chain plant will pose problems in containing the radiation at the entrance and exit of any microwave oven, but with a batch plant, where it is usual to use a sealed oven, the possibilities of microwave heating are much more feasible.

Control of the drying and vulcanising temperatures by monitoring the film temperature *via* infrared thermometers is now relatively simple, and can give much more accurate control and avoid the problems caused by over- or underdrying.

It is possible to buy temperature loggers that can be taken through the production sequence, and these loggers will accurately record the temperature through every step of the production process. In this way, the evenness and accuracy of the heating can be verified throughout the whole dipping process and any 'dead' heating elements quickly detected.

17.3 Possible new materials

17.3.1 New latices

There are other types of naturally occurring rubber, apart from *Hevea* rubber. Other species of plant produce *cis*-1,4-polyisoprene, or 'natural' rubber, although few of these offer any hope of commercial exploitation. Two that do, however, are the guayule shrub and the Russian dandelion. There has been considerable interest in guayule for many years, and latex from this plant is now commercially available and generating considerable interest in certain quarters. Guayule latex contains proteins, but these proteins are different from those found in *Hevea* latex and have not been found to cause allergenic reactions, as touched upon in Chapter 16. Guayule latex is not a direct alternative to *Hevea* latex however, and special compounding and formulation procedures are required to obtain satisfactory dippability and physical properties. Rubber derived from the latex of the Russian dandelion was discovered back in the 1930s, but was never regarded as a serious alternative to the rubber tree. However, as with guayule, considerable recent development work and investment has advanced this plant as a potential commercial source of rubber latex for dipping.

17.3.2 Fillers

One of the more attractive ways of improving performance is through reinforcing fillers. However, as noted in Chapter 5, there are particular problems in reinforcing latex systems. The protective layers on the surface of the particles, that confer stability to the system, can also inhibit intimate contact with any filler particles, and finely divided fillers may also rob the latex particles of stabiliser and promote coalescence. However, interest in the use of fillers in latex system continues, and at the time of writing, work is underway exploring the use of graphene as a filler for latex products, especially condoms. Graphene is a two-dimensional allotrope of carbon that is only one atom thick. It possesses many interesting properties, including incredible strength; however, as mentioned above, the effect of interaction with the stabiliser systems in a latex system could reduce any reinforcement. It will be interesting to see what benefits this unique material can offer the latex dipping industry.

17.3.3 Nanotechnology

The emergence of nanotechnology – working with materials with at least one dimension sized from 1 to 100 nanometres (0.000001 to 0.0001 mm) has the potential to offer advantages within the latex dipping industry. Nanofillers, some of which are described in Chapter 5, may offer increased performance when incorporated into dipped latex articles, although the enormous surface area of such materials will have the potential to rob the latex of stabiliser and could pose problems. On the other hand, presentation of compounding ingredients with such a high surface area may enhance their effectiveness, and as mentioned below, the use of nanosized zinc oxide (ZnO) is claimed to allow lower levels of zinc to be used in a NR formulation. An awareness of progress in nanomaterials is recommended to those in the latex dipping industry.

17.4 Increasing regulatory requirements

Many of the products made by latex dipping are medical devices, and as such, are regulated. This regulation is tending to increase, especially if significant medical claims are made, and any new medical device faces significant hurdles to overcome before it can be put on the market; these can include extensive clinical trials, toxicity testing and licensing. These constraints place restrictions on the development of new materials and technologies. The bar to commercialisation is raised, and new developments must demonstrate increasing technical and commercial advantages to be pursued. However, whilst these regulations may be seen to inhibit innovation, they can also act to help protect any truly innovative developments, by making it more difficult to copy them. Regulation can be viewed as a friend, as well as a burden.

17.5 Increasing environmental requirements

Like regulatory requirements, environmental constraints are increasing. Zinc in effluent, in particular, is being targeted as damaging to the environment and there is ongoing research into alternative activators for latex systems. Other metal oxides have been evaluated as a replacement for ZnO, but none so far has been found that equals the zinc salt. The best appear to be oxides of cadmium, lead and mercury, but the environmental benefit of replacing zinc with these materials appears to be dubious. The so-called ‘multifunctional additives’ (amine-bridged amides) have been used to enhance the cure of CR in dry rubber systems [1], but their applicability to latices has yet to be fully explored. Perhaps a more fruitful approach is reducing the concentration of ZnO in the formulation, and to this end the use of ‘nano’ ZnO shows promise. The nanomaterial has an average particle diameter of around 30 nm, compared with the approximately 0.6 μm of standard ‘red seal’ ZnO and 0.2 μm of the ‘active’ ZnO.

Another possible way of reducing the zinc concentration in the formulation is the use of zinc complexes. One method that offers potential is the use of zinc complexed with clay. Montmorillonite clay has many sites that can react with cations and it is possible, through an ion-exchange mechanism, to form a complex of zinc with the clay. This complex has been shown to be active in the vulcanisation of NR latex, giving acceptable physical properties, under the conditions employed in the study [2]. It is to be expected that work will continue to reduce the environmental impact of latex dipping and there appear to be several avenues to explore in this aim. The importance of manufacturing activities on the environment is not expected to diminish: rather the opposite.

17.6 Drug delivery

Some polymers can be successfully used for drug delivery and latex dipped articles are no exception. There are already, at the time of writing, condoms that can deliver drugs, such as spermicides, local anaesthetics or erection-enhancing chemicals through a lubricant or gel applied to the condom, but it is possible in the future that active materials can be incorporated into the rubber film itself for delivery to the user or outside world. This would be valuable in condoms for releasing drugs to help combat the risk of human immunodeficiency virus/acquired immune deficiency syndrome, to act as a spermicide or to help erectile problems. In this scenario, surgical gloves could be formulated with active materials to help maintain sterility and prevent the transfer of pathogenic materials if the glove is punctured. Houseware gloves could be loaded with emollient creams to improve the protection to hands, or microbicides to keep the glove 'fresh'. These are ideas that have been investigated in the past, but with the continuing emergence of new technologies and materials, future development could prove more fruitful.

17.7 Specialised coatings

There are products available with specialised surface coatings. For example, the Biogel® surgical glove, carrying a special low friction internal coating, was the first surgical glove able to be donned without the use of additional lubricating powder. These coatings are also able to modify the surface of the dipped latex article to confer particular properties, and may also be able to be used to deliver drugs and materials, as discussed above; but one major problem to be addressed with coatings is that of adhesion. Most dipped latex articles are very elastic and will be stretched in use or upon donning. For a coating to be effective, it must be compatible with the rubber surface and persistent during normal cycles of use. This may be less of a problem with single-use items such as surgical gloves or condoms, but is more difficult for multiple-use products. Nevertheless, there has been significant progress in coatings since the Biogel® glove was introduced and this area of development could be very interesting.

17.8 New materials

Many of the products made by latex dipping could also, in theory, be made by other materials. However, the advantages of dipping a latex were outlined in Chapter 1 – namely the convenience of a water-based system operating at a moderately high total solids content. Polyurethanes (PU) can be very strong, elastic materials, ideal for many product applications in latex dipping. There have been several attempts to develop PU latices for dipping, some of which have shown promise. However, at the time of writing it is thought that there are no PU latices being used for dipped latex products. Owing to the wide variety of physical properties that can be obtained from PU, this is an area that is available for exploitation by the right materials. Other elastomers have been explored, including those prepared as artificial latices, such as isoprene-isobutylene (butyl) rubber and chlorosulfonated polyethylene (Hypalon), but it is considered that the versatility of PU offers the most intriguing possibilities in the use of new materials.

17.9 Reducing costs

As with any other manufacturing operation, there are always commercial pressures on latex dipping. These will not ease in the future. So what can the future bring to help manufacturers maintain or reduce costs? The best way to reduce costs is to improve quality. Improving quality brings many benefits, including higher yields, leading to lower unit labour, material and energy costs. Lower levels of scrap will reduce waste disposal costs and offer the opportunity to reduce testing costs. In addition, there could be an improved perception of the company by customers.

Other ways in which costs may be reduced could be by taking advantage of any improvements in areas such as raw materials, production processes and automation. The rate of progress in many areas is impressive and difficult to follow, and it is recommended that manufacturers maintain a watching brief on the developments in the latex dipping and wider polymer and material areas. However, as discussed above, looking at the, perhaps 'lower-tech', quality improvement route will often prove to be the most useful way to reduce costs.

17.10 Improving measurement processes

17.10.1 Measuring the extent of prevulcanisation

Many of the measurement processes used in latex technology are well established and need no improvements. The same cannot be said of the measurement of prevulcanisation. Most natural rubber latices are prevulcanised before use, and other latices such as polychloroprene, acrylonitrile-butadiene and synthetic *cis*-1,4-polyisoprene are, if not prevulcanised, allowed to mature for some time after compounding. In order to ensure that the latex is in the correct state for dipping, it is customary to measure the extent to which any prevulcanisation has occurred. Too much prevulcanisation may result in the finished product having poor physical properties, whilst too little prevulcanisation can give problems on the dipping line (e.g., difficult stripping or stripping creases) as well as possibly affecting the physical properties. However, there is not yet a quick, easy and, most of all, accurate measurement of the degree of prevulcanisation. The various methods in use at present are described in Chapter 2, but all are considered to have drawbacks. Prevulcanisation does not interfere with the properties of the latex as a colloidal dispersion, so colloidal properties are no help in measuring the extent of prevulcanisation. A prevulcanised latex looks pretty much the same as an unprevulcanised latex, so visual measurements are no good. The density of NR changes slightly with prevulcanisation, but the change is too small to be of any use as a method for monitoring.

The ideal test would be something like a pH meter – you take a probe, immerse it in the latex and read the answer off the display. The development of such a test would not revolutionise the industry, but would assist in improving the quality and reliability of the manufacturing process. A project for the future.

17.10.2 Predicting the 'dippability' of latices

There can be occasions when a new batch of latex arrives with satisfactory specification, and results from goods inwards testing are satisfactory, but, when compounded, the latex does not perform as expected on the dipping line. Here is where it would be good to have a test that would predict, accurately and reliably, how a latex would behave when compounded to the standard formulation and put on the dipping line.

There has been a lot of work over the years to try and develop a test for latex 'dippability' or 'processability', but none yet has been found to be reliable in all circumstances. This problem is considered to revolve around stability: sufficient stability to be stored, handled, pumped and agitated, yet sufficiently unstable to form a coherent film quickly and easily when dipped. There has been a lot of work on this problem over the years, and there are tests available to estimate mechanical stability under various conditions, with and without agitation, heat, the addition of destabilising chemicals and so on. However, none has yet been found to give a reliable prediction, in all situations, of how a latex will perform on the dipping line. As with the measurement of prevulcanisation, the development of such a test would not revolutionise the industry, but would assist in improving the quality and reliability of the manufacturing process.

References

1. C. Hepburn, *Rubber Technology International*, 1985, **24**, 5.
2. G. Heideman in *Reduced Zinc Oxide Levels in Sulphur Vulcanisation of Rubber Compounds; Mechanistic Aspects of the Role of Activators and Multifunctional Additives*, University of Twente, Enschede, The Netherlands, 2004 [PhD Thesis].

Abbreviations

3D	Three-dimensional
ACN	Acrylonitrile
AIDS	Acquired immune deficiency syndrome
BA	Butyl acrylate
BHT	Butylated hydroxytoluene
CB	Carbon black
COA	Certificate of analysis
Cp	Capability index (process mean centred between specification limits)
Cpk	Capability index (process mean not centred between specification limits)
CR	Polychloroprene
CSR	Chlorosulfonated polyethylene rubber
CTQ	Critical to quality
DC	Double-centrifuged
DNPD	*N,N'*-di-2-naphthyl-*p*-phenylenediamine
DPG	*N,N'*-diphenylguanidine
DPPD	*N,N'*-diphenyl-*p*-phenylenediamine
DPTU	*N,N'*-diphenylthiourea
DQ	Design qualification
EPDM	Ethylene-propylene diene monomer rubber EV Efficient vulcanisation
FDA	US Food and Drug Administration
HA	High ammonia
HIV	Human immunodeficiency virus
IIR	Isoprene-isobutylene rubber
IQ	Installation qualification
IR	Synthetic *cis*-1,4-polyisoprene rubber
ISO	International Organization for Standardization LA Low ammonia
LSL	Lower specification limit
MAA	Methacrylic acid
MBI	Mercaptobenzimidazole
MBT	2-Mercaptobenzothiazole
MFFT	Minimum film-forming temperature
MMA	Methyl methacrylate
MMT	Montmorillonite
MSDS	Material safety data sheet
MST	Mechanical stability time
MW	Molecular weight(s)
Na-MMT	Sodium montmorillonite
NBR	Acrylonitrile-butadiene rubber/nitrile rubber(s) NR Natural rubber
OQ	Operational qualification
PAN	Polyacrylonitrile
PMMA	Polymethyl methacrylate
PQ	Process qualification
PS	Polystyrene
PU	Polyurethane(s)
PVC	Polyvinyl chloride
RPI	Relative precision index
RT	Room temperature

https://doi.org/10.1515/9783110638097-018

SBR	Styrene-butadiene rubber
SD	Standard deviation
Semi-EV	Semi-efficient vulcanisation
SIX	Sodium isopropyl xanthate
SMBT	Sodium 2-mercaptobenzothiazole
SPC	Statistical process control
T_g	Glass transition temperature
TSC	Total solids concentration
USL	Upper specification limit
UV	Ultraviolet
XNBR	Carboxylated nitrile rubber
ZDBC	Zinc di-*n*-butyldithiocarbamate
ZDEC	Zinc diethyldithiocarbamate
ZIX	Zinc diisopropyl xanthate
ZMBI	Zinc mercaptobenzimidazole
ZMBT	Zinc 2-mercaptobenzothiazole
ZMTI	Zinc 2-mercaptotoluimidazole
ZNBX	Zinc di-*n*-butyl xanthate

Index

https://doi.org/10.1515/9783110638097-019

Bei Fragen zur Produktsicherheit wenden Sie sich bitte an:
If you have any questions regarding product safety,
please contact:

Walter de Gruyter GmbH
Genthiner Straße 13
10785 Berlin
productsafety@degruyterbrill.com